COP TALK!

*How To Listen In On
Law Enforcement Communications*

By Laura E. Quarantiello

TIARE PUBLICATIONS

Copyright © 1999 by Tiare Publications

All rights reserved. No part of this book may be reproduced or transmitted in any form or by any means electronic or mechanical, including photocopying, recording or by information storage or retrieval systems, without permission in writing from the publisher, except by a reviewer who may quote brief passages in a review.

Cover photo courtesy of National Communications Magazine/ Bearcat Radio Club

Book layout by John C. Herkimer,
Next Wave Graphics,
Caledonia, NY

2nd Edition

Published by Tiare Publications,
P.O. Box 493, Lake Geneva, WI 53147
http://www.tiare.com

ISBN: 9-936653-79-5
Printed in the United States of America

Library of Congress Cataloging-in-Publication Data

```
Quarantiello, Laura E., 1968-
    Cop talk! : how to listen in on police radio communications / by
  Laura E. Quarantiello.
       p.   cm.
    ISBN 0-936653-79-5 (paper)
    1. Police communication systems.  2. Radio--Monitoring receivers.
  I. Title.
  HV7936.C8Q37  1998
  363.2'4--dc21                                            97-35220
                                                              CIP
```

TABLE OF CONTENTS

PREFACE TO THE SECOND EDITION 7
INTRODUCTION 9
1. THE WORLD OF LAW ENFORCEMENT COMMUNICATIONS 13
    Why Listen 14
    What Will You Hear? 14
2. A RADIO PRIMER 15
3. A SCANNING PRIMER 17
    Strike Up the Bands 19
    Frequencies 19
    Reception 20
    Equipment for Monitoring 21
    How to Set Up Your Scanner Frequency Banks 23
    Two-Scanner Monitoring 24
4. FINDING LAW ENFORCEMENT FREQUENCIES 25
    Resources for Frequencies 25
        Books 25
        Magazines 26
        Clubs 26
        Other Monitors 26
        Radio Retail Outlets 26
        The Internet 27
        Police Departments 27
    Finding Frequencies With Your Scanner 27
    Frequency Counters 28
5. TECHNICALITIES OF POLICE RADIO SYSTEMS 29
    Repeaters and Simplex 29
    Vehicular Extenders 30
    Talkaround 30
    High Tech Police Radio 30
    800 MHz Trunked Systems 30
    Coping With Encryption 32
    Computer Aided Dispatch and Mobile Data Terminals 33

6.	RADIO JARGON 35
	Police Phonetic Alphabet 35
	Police Incident and Disposition Codes 36
	Cop Shorthand: Abbreviations in Communications 37
	Unit Numbers: How to Identify Who's Who 39
	Radio Units 40
	Uniformed Division 40
	Investigations Division 41
	Special Units 41
	Keeping Track by Keeping Notes 42
7.	MONITORING LAW ENFORCEMENT COMMUNICATIONS 43
	Dispatch 43
	Inquiry 44
	Tactical 44
	Car-To-Car 46
	Emergency 46
	Other Specialized Frequencies 47
	Frequency Numbering 47
	Cellular Telephones 48
	Undercover Operations 49
	Emergency Situations 50
	Techniques for Low-Profile Mobile Monitoring 50
	Monitoring on Foot 51
8.	THE LAW - ON MONITORING THE LAW 53
	How to Find Out if Mobile Monitoring is Legal in Your State 53
	Responding to a Call: The Line Between
	On-Looking and Interfering 54
9.	HOW YOU CAN HELP LAW ENFORCEMENT 57
10.	A BADGE 61
11.	IN CLOSING 65
12.	APPENDIX 67
	The Miranda Warning 67
	Search Warrant Format 69
	Glossary 71
	Commonly Used Abbreviations 77
	Law Enforcement Radio Frequency Ranges 79
	Nationwide Police Common Frequencies 79
	Tracking and Surveillance Frequencies 81
	Police Radar Frequencies 81
	General Frequency Band Allocations 83
	Sample Police Codes 87

DISCLAIMER

The purpose of this book is to provide information about law enforcement radio listening that will give the reader a greater understanding and enjoyment of police communications. The information in this book is presented for educational purposes only and is intended for legal use only. Neither the author, the publisher nor their sales agents accept responsibility of any kind for any illegal usage or applications which might be made of this information.

The reader is advised and encouraged to practice responsible scanning, which includes not visiting crime or accident scenes, not interfering with law enforcement operations and not repeating or distributing what is heard to anyone who is not a party to the original transmission. The reader should become familiar with the federal and state or local laws which may apply to the use of radio scanners and law enforcement monitoring.

PREFACE TO THE SECOND EDITION

Law enforcement communications have undergone profound change since the first edition of *Cop Talk* was published. Today, 800 MegaHertz trunked radio systems are common, replacing the single frequency dispatch systems we've long been used to monitoring. Mobile Data Terminals (MDT) place the power to conduct license, registration, and identity searches at the fingertips of police officers, instead of having to run the information through a dispatcher. Automatic Vehicle Locators (AVL) hook police vehicles into the headquarters computer, allowing dispatchers to track each individual unit's location on a map, while devices such as LoJack make finding stolen vehicles easier. Technology has arrived in full force and its effect on monitoring law enforcement communications is undeniable.

Not to be outdone however, the communications industry and radio hobbyists have responded by introducing computer-controlled scanning interfaces, trunk tracking scanners, and computer software packages to make monitoring easier and more efficient. Also available are frequency counters/decoders, wide coverage receivers and encryption descramblers. Monitoring police communications has never been easier or, at the same time, more involved than it is today. But, although the tools have improved and new techniques for listening have been introduced, the basic foundations remain: in the end it's just you and your scanner radio.

Cop Talk was born as a result of countless conversations with scanner listeners whose main interest was law enforcement radio. "What are they saying?" and "How can I understand it?" were common questions. This book provides the answers, though even with the inside information given here you will still have to sit down and listen intently. There is a lot of information out there which you simply have to obtain on your own. There's no other way. Even after you've had a lot of experience there will be times when you'll still puzzle over what you hear. No two law enforcement agencies use the same codes, abbrevia-

tions, call signs, frequency designations, unit assignments - all are individual to each department and require effort on your part in order to figure out what they mean. *Cop Talk* will give you the basics, as well as numerous tips and tricks for making your listening sessions easier, but the real work, and any rewards you receive, are up to you.

INTRODUCTION

It is early morning, just past six a.m. as "A" shift deputies begin to check in on Frequency Two. It is a familiar litany to me, as unit after unit comes on the air with call sign, ARJIS (Automated Regional Justice Information System) number, and unit ID number. Interspersed are the weary voices of "C" shift deputies calling 10-7 as they pull into the Sheriff's Station and end their 11 to 7 tour.

The dispatcher, whose name I don't know but whose voice is as recognizable as my own, begins to hand out the usual run of 459 (burglary), 594 (vandalism) and 10851 (stolen vehicle) calls that make up the bulk of the work in the morning for day shift units. The first traffic unit to check in, fresh out of the station parking lot, is given a three car collision on Highway 78, with medics responding. 11-49's are popular numbers this morning as deputies begin to stop drivers bending the laws in their attempts to make it to work on time. The Inquiry frequency begins a non-stop chatter as license and registration numbers are run through the computer for wants, warrants, suspensions or anything else that might prompt a deputy to write a citation, make an arrest, or impound a vehicle.

By ten a.m. the routine calls are dwindling and other calls are pending all across the city, some of them 415's. Domestic disputes are one of a cop's worst nightmares, representing an unknown, potential danger as a badge and uniform attempt to keep the peace in household quarrels. Many sheriff's units here are one-man. Some deputies choose to handle the 415's solo, while others request backup. A 211 (armed robbery) interrupts things and draws deputies away for awhile, as do a series of higher priority calls. By noon, deputies earnestly begin seeking lunch breaks with "is the beat clear for Code 7?" The dispatcher does her best to clear units to lunch sequentially, but there are never enough deputies on the street to handle everything and some calls inevitably become delayed. High priority calls go out immediately, while others are held until units can be cleared to handle them.

Frequency One, the sheriff's Car-to-Car channel, keeps me tuned in on inter-deputy chatter. Two units are searching for an address and their frustration in being unable to find it comes through on the radio. Finally, someone straightens out the map book and the units call 10-97 (on scene).

Shortly after two p.m., a 245 (assault with a deadly weapon) call is dispatched and three units roll. Five minutes later, another 245 is given out for a location three blocks from the first. Sensing a pattern here units break from their other calls and before long a string of 417 (person with weapon) calls have flooded the dispatcher. She struggles to relay all of the incoming information to the responding deputies. There is slight confusion as to the suspect's description, but eventually the reports are collated. 90 Paul Sam, the patrol sergeant, organizes his deputies and sends them to Tactical Frequency Nine.

The 245 suspect and the 417 suspect turn out to be one and the same. Two gunshot victims lay awaiting paramedics and suddenly we have more than a routine afternoon going on here. Amid the confusion of "B" shift coming on duty at 3 p.m. and "A" shift trying to clear out of service, there are rapid fire radio calls. The dispatcher holds it all together somehow, simultaneously working radio and computer, keeping track of unit locations and assignments.

Over on Frequency Nine the patrol sergeant has quickly set up his deputies to form a perimeter. Hopefully, if they have moved fast enough, they have the suspect penned in somewhere between the perimeter units. 90 Paul Sam calls for a K-9 unit to do an area search. ASTREA, the sheriff's helicopter, reports in only thirty seconds away and will be a welcome addition. The Administration channel, Frequency Six, relays victim information to the patrol sergeant - one man is dead on arrival at the hospital. Our bad guy is now a 187 (murder) suspect.

The bulk of the action is on Dispatch Frequency Three, Tactical Frequency Nine, and Car-to-Car Frequency One. Two K-9's call on-scene, with another coming, and the search becomes one of feet, paws, and rotary wings. Some "A" shift units are still on duty and will remain on overtime until relief is available or the suspect is arrested. Who could have imagined this morning what the afternoon would bring?

The search is slow but methodical, with Dispatch advising of one additional 417 call in the midst of it all, giving deputies a direction. The hunt narrows. Deputies on foot are using handheld radios and their voices strain and break as they talk, walk, and run. In the background I can sometimes hear the bark of K-9's and the rattle of helicopter rotor blades.

The scanner keeps me right on the scene, in the middle of the action as, in rapid succession, I join a patrol unit, a K-9 deputy on foot, two deputies in the air aboard a helicopter, and "peer over the shoulder" of a dispatcher. I congratulate myself for my smart pre-planning in the way I set up the scanner's frequency bank. Following the action is easy. I'm hearing everything.

Suddenly, just when it looks as if the bad guy has gotten away, that the hard work and worn shoe leather has been for nothing; as the search seems to be dragging and I'm thinking about leaving the radio:

"90 Paul 5, I have the suspect running northbound West Lake Street. Have a unit cover the north end!"

Voices are everywhere; every frequency is alive. Sirens yelp in the background, dispatchers keep a running commentary on the direction of travel. Like a fox who has broken from cover and run, the hounds are after him. The intensity level increases by several degrees.

"Paul 5, taking the suspect down!"

Down? What does he mean? One fractured transmission and the frequency goes quiet. Has someone been shot? Is the suspect in custody? This is a tense moment for any scanner listener, when information is an airwave away, but only silence comes from the radio. Pictures of what might be happening out there flash across my imagination. I wait. The scanner hums. It, too, is waiting.

"90 Paul Sam, suspect in custody, all units maintain the perimeter."

In custody! What a relief! Sweaty, parched and stiff from tension I lean back and listen to the wrap up. Units are all accounted for, relieved to return to the station or their beats. Voices lose their strain and take on a weary calm. The dispatcher begins to hand out calls that have piled up while units were engaged in the search. Before long it sounds like a routine evening out there again. But for how long?

I've been a law enforcement communications monitor from the day I bought my first scanner. Every time I think I've heard all there is to hear something new and strange will happen on the radio. That's the reason I listen - to gain an insight into what's going on in my city; the good, the bad - and the weird! Through it all I never fail to come away from the radio without a sharp respect for the people who put their lives in jeopardy to keep me safe.

The information in this book comes from my own personal experiences as a scanner listener, and my research as a writer on communications topics. Everything you read here has come from public sources and is just as available to you as it is to me. You would be

amazed at the amount of information about law enforcement available from library reference sections, city records, and publicly available sources, not to mention what you can dig up through your own research and perseverance.

My thanks go out to all of those who assisted me and reviewed this manuscript: law enforcement officers, dispatchers, and fellow listeners who gave unselfishly of their time and knowledge so that others may gain insight into the police profession. Any errors herein are the result of my own inattention and the opinions expressed are strictly mine, unless otherwise indicated.

There are no adequate words to express my feelings to the men and women of the San Diego County Sheriff's Department Station 90, where I work as a volunteer. Thank you for allowing me to be a part of your world. This book is dedicated to all of you.

- Laura Quarantiello
lauraq@funtv.com

1

THE WORLD OF LAW ENFORCEMENT COMMUNICATIONS

The first police officers summoned assistance by rapping on manhole covers with their nightsticks. It was primitive communication to say the least, but it was effective. As police work evolved into a profession and technology improved, telephone call boxes positioned on street corners served as the officer's link with headquarters. Every beat officer carried a special key which would open the locked boxes. The radio system that replaced these telephones was used only by high-ranking officers at first, then gradually found its way into patrol cars and eventually into the handheld units every cop wears on his equipment belt today.

The two-way radio is a mandatory piece of equipment carried by all officers in the field. Each patrol unit usually has a radio mounted permanently within it, and each officer usually carries a handheld radio with him wherever he goes. Handhelds are so important they are each given a property number by the department. They are recharged after every shift so they're ready to go at all times. The radio is an officer's connection to headquarters, to license and registration checks, wants and warrants information. But most of all it is his link to assistance. Without a radio an officer's call for help would not carry very far. The streets of today all but demand that an officer not be left out in the cold, away from back-up. The radio is the lifeline.

The phrase "law enforcement communications" encompasses all radio transmissions by local, state, and federal services such as sheriff, city police, state police, highway patrol, marshal, and federal agencies, such as the Federal Bureau of Investigation, Bureau of Alcohol, Tobacco and Firearms, and the Drug Enforcement Administration. The Corrections and Probation departments could also be considered under this heading.

Police communications vary widely, from the mundane to the exciting: patrol division dispatching, special investigations, narcotics

division operations, gang enforcement and undercover operations, as well as traffic control and search and rescue coordination. If it involves keeping the peace and protecting the public, it probably falls under the jurisdiction of the police.

WHY LISTEN?

Each listener can probably given you a dozen different reasons why police communications occupy center stage on their scanner. The fun of it, the interest in the job, wanting to know what's going on, keeping track of crime in their area and being informed.

Your reason might mirror these, or it might be totally different, but it cannot be denied that the ability to peek behind the badge and listen in on law enforcement is a pursuit shared and enjoyed by most scanner owners. Police monitoring is the most popular form of scanner listening and the primary reason why most people first purchase a scanner. Whatever your reason, police radio has an allure all its own.

WHAT WILL YOU HEAR?

Police radio communications deal with everything and anything related to crime and the enforcement of city, state and federal laws. You will hear patrol cops responding to audible and silent alarms, domestic disturbances, trespassers, fights, assaults and shootings; traffic officers ticketing moving violators, responding to accidents, hit and runs, and road hazards; detectives investigating all manner of crimes, serving warrants, searching for suspects, following up on leads; narcotics officers on stake-outs and drug busts; police helicopters conducting patrols, and search and rescue operations. The question really isn't what you'll hear, but what won't you hear!

> *"When people observed me in my uniform, on the street, they saw the shirt and badge, a faceless and heartless representative of authority. But, like all cops, I was in there, flesh and blood and heart. They are in there, behind what is often mistakenly called a shield. They hear you and see you, and know and feel the feelings you experience because they feel them, too."*
>
> - Sgt. C.P. McDonald (Retired)
> Fort Lauderdale, FL Police Department
> "Blue Truth", 1991

2

A RADIO PRIMER

Take a look at any AM/FM radio you may have handy. Notice that the dial shows a series of numbers from 540 to 1600 or 1610 (1700 on newer sets) and from 88 through 108. These are the AM and FM broadcast radio bands which most of us are intimately familiar with, which stare out at us from stereos, portable boom boxes and car radios.

Everything from a siren, a ringing telephone, a Mack truck rumbling by or the voice of a friend causes the vibration of air molecules. These vibrations form invisible pressure waves that radiate out and strike our eardrums. One wave is equal to one cycle. The closer together the waves are, the faster the vibrations occur and the higher their tone becomes. Once the frequency of the vibrations reaches about fifteen (cycles) per second, they become audible to the human ear, which can hear audio frequencies up to almost 20,000 cycles (20 kHz) per second, at which point the frequency exceeds the audible range and is known as ultrasonic.

We can thank German physicist Heinrich Hertz for discovering the wave properties of electromagnetic energy. Hertz, a guy who spent a lot of time looking for a connection that no one else had ever even thought about, proved that electricity can be transmitted in waves, which travel at the speed of light and possess many of the same properties as light. His experiments helped pave the way toward the development of radio.

Radio signals, it was discovered, are composed of repetitive waves of electrical energy. The frequency of these waves is rather high - into the thousands and millions of cycles per second. The technically inclined tell us that 1000 cycles per second equals one kilocycle and one million cycles per second is the same as one megacycle. To express the frequency of a radio signal we used kilocycles and megacycles. Some time ago, however, it was decided to honor Heinrich Hertz by using "Hertz" instead of "cycle." Of course, this changed everything all up the line, turning kilocycles into kiloHertz (kHz) and megacycles into MegaHertz (MHz). Just when we thought we knew where to find our

favorite radio station - on 1170 kilocycles or 100.1 megacycles, for instance, we had to start saying 1170 kiloHertz and 100.1 MegaHertz instead. Now, of course, radio broadcasters use cute names like "Flash 100" and "Mighty 690," proving that we don't even need to use kiloHertz and MegaHertz anymore...or do we?

Look at the ends of your AM/FM dial and you will see that the numbers seem to have a finite limit. That is, they end at 1600 or 1700 kHz on AM, and 108 MHz on FM. These numbers, however, are the boundaries of just two tiny slices of the electromagnetic spectrum; they are far, far from the be all and end all of radio.

If it were possible to look beyond the end of the dial and get past the artificial tuning limits the manufacturer built into your radio, you would discover that the entire radio portion of the spectrum is vastly larger than the AM and FM bands; a solar system compared to a galaxy. The radio spectrum begins at around 50 kHz and extends up to and beyond 20,000 MHz.

The segments where police, fire, EMS, railroads, airplanes and the like are found is in a range beginning at 30 MHz and running to about 950 MHz. You won't be able to tune these frequencies on your AM/FM stereo, however. You'll need the services of a special radio for that.

"Every time I am convinced that I should quit today and go sell tee-shirts on the beach, something happens and the blood gets racing and I realize that I have more to learn and do. With two grandkids and one on the way, I should have enough (even my husband reminds me that I'm not that young and should be off the street - but he told me that when I was in my twenties!) Heck, I can't even quit smoking and this addiction is a lot more fun. I'll probably be fired long before I quit. As sad as it is out there, w are still accomplishing something, or at least holding that thin blue line."

- Sgt. Eva Staunton
Gwinnett County, GA Police Department

3

A SCANNING PRIMER

Webster's Dictionary carries the best definition of scanning that I've ever found: "to examine stored material for a purpose; to examine channels for input/output activity." Unfortunately, as accurate as it is, Mr. Webster's explanation applies to a broad range of activities, from quickly perusing a newspaper's front page to channel-surfing on your TV. You can scan a page, scan a crowd and scan the refrigerator shelves. In effect, we scan things every day. But "real" scanning, the term given to the activity which includes police communications monitoring, involves radio. If you've ever twirled the dial on your AM/FM radio, spinning past news, sports reports, talk shows and music of all kinds as you searched for a certain station or just something interesting to listen to, then you've done radio scanning at its most basic. If Webster had known about this kind of scanning he probably would have added another section to his definition: "the use of a special radio receiver called a scanner to tune into the communications of non-commercial, non-broadcast services such as police and fire departments." A scanner is able to search the radio frequency spectrum, listening for active transmissions, stopping when it hears voices, and then continuing its search when that frequency falls quiet.

A scanner can take you from the shotgun seat of a police patrol car to the top rung of a 75-foot fire department ladder to the cockpit of an F-14 Tomcat pulling eight g's off the deck. Turn a knob, punch in some numbers and you're off; probing the airwaves, listening to some of the thousands of voices constantly in the space around us. From routine calls to a four alarm fire - in seconds you're on-scene without leaving the comfort of your chair. Forget Top 40, forget talk radio's whiners and complainers. With a scanner your next easy listening station can be the airport control tower or the county sheriff's dispatcher. Number one on today's countdown might be a major accident or a stolen vehicle pursuit. You never know what you'll hear. Indeed, scanning is much more than a dictionary's dry definition. It's an intriguing, exciting pastime; a world all its own.

THE BIRTH OF SCANNING

Some brilliant mind eventually came to the realization that humans are insatiably curious beings who were probably growing tired of plain old AM and FM broadcasts. After all, there's an entire radio spectrum out there! So the idea was born of producing a radio that did more than tune the AM/FM bands. This radio's dial covered the "hidden" bands, those that didn't appear on ordinary receivers, those where the two-way communications of police, fire and other services could be found. You tuned this radio much like any other, with a knob that allowed you to move up and down the dial. It required a light and delicate touch to accurately find a station, and all tuning was manual: once you stopped turning the dial, the radio stayed on that frequency.

Well, this was a fine and wonderful invention and before long listeners were tuning all over the place, discovering communications they hadn't been aware of before. In time, they found favorite frequencies which they returned to again and again. Maybe it was the local police or fire department channel, or perhaps that of the airport control tower. But you could only listen to one of these at a time and there were often long stretches of silence during which you wondered what might be happening on the other frequencies. If you decided to find out, you had to retune and find each station anew every time you wanted to hear something different. This led to some people placing small pencil marks and notations on the dial face of the radio, a sin that purists still consider something quite close to sacrilege. It may have been one of those purists who, perhaps with a sense of outrage, declared "well, fine, if you don't like tuning back and forth trying to find frequencies, I'll just give you a radio that you can load your favorite frequencies into and it'll tune them for you automatically." So was born the scanner.

From a manually-tuned radio, the scanner evolved into a device that you could load with crystals for each frequency you wanted to monitor. From there, things leaped ahead technologically, giving us the modern scanner that can tune, remember and search through any combination or band of frequencies. The one thing that hasn't changed through all these advancements is the bread and butter of radio listening: the frequency. Scanner listeners still live and breathe for numbers, just like those we'll talk about throughout this book. Understanding those numbers is the key to getting the most out of this book.

STRIKE UP THE BANDS

The radio spectrum we've been talking about is divided into individual bands (groups) of frequencies. The standard AM broadcast band where you catch the ball game or the latest stock market information is known variously as the MF or Medium Frequency band or AM broadcast band. The FM broadcast band, on the other hand, falls into an area known as the VHF or Very High Frequency band. The major bands of specific interest to scanner listeners are the following:

    30.000 - 50.000 MHz   VHF-Low Band
    138.000 - 174.000 MHz   VHF-High Band
    406.000 - 512.000 MHz   UHF (Ultra) High Frequency Band

Any scanner you buy should cover these basic bands as a matter of routine. More advanced and expensive scanners add other frequency ranges in the spectrum such as the 108-137 MHz civilian aeronautical bands, 225-400 MHz military air band and the 800 MHz public safety and business band. Top-of-the-line scanners that cover 25-1300 MHz and higher are often billed as continuous coverage

FREQUENCIES

Within each band are individual frequencies, expressed in much the same way as those in the AM/FM broadcast bands, only with greater accuracy. When giving a scanner frequency it's not enough to say that a station is operating on *about* 154 MHz. Unless it's exactly on 154.000 MHz. you'll have to be more specific. There could be a station licensed on 154.000 MHz, one on 154.015 MHz, another on 154.030 MHz and so on. A few kiloHertz either way does matter here. Frequencies are spaced apart by a certain number of kiloHertz which differs depending on the band, commonly in 5, 10, 12.5, or 30 kHz steps. Spacing is a hot topic in the radio community lately, as the Federal Communications Commission is slowly implementing a narrower channel spacing scheme for stations in the public safety services. This is expected to be complete and in place within ten years, if all goes according to plan.

By the way, the decimal point in the above examples is commonly pronounced as "point," rather than "dot." Hence, 163.125 MHz would be spoken as "one six three point one two five MegaHertz." Some scanner listeners break protocol and eliminate the point when speaking of a frequency, but you'll always see the decimal used when frequencies are written out, as well as in your scanner's display window. Some

scanners show frequencies to four digits after the decimal point, meaning that a typical display might read 506.7875 MHz. If your scanner doesn't display four digits to the right of the decimal, and you see something like 506.785 MHz when you try to enter the full frequency, don't panic, the scanner has just rounded off the number. You're still close enough to the actual frequency to hear the transmissions.

RECEPTION

Radio waves are fascinating and finicky things, especially those in the VHF/UHF ranges. They are easily diverted, weakened or absorbed. Once they escape the transmitter and antenna they can get lost in hills, valleys and other quirks of terrain and architecture. Depending on the power of the transmitter, a radio signal may have an easy time making it to your scanner or it may dwindle away to nothing while still within several hundred yards of its origin. Reception of radio signals depends on many different factors, such as your location, the quality of your scanner and antenna, what type of building you're in, your height above the horizon, the location and power of the transmitter, and more. The best tool we have to compensate for this (some might say the only one) is our antenna.

The antenna that came with your base or handheld scanner is basically inadequate for anything but general listening. It'll do fine if all you need to hear are nearby high-powered stations, but if you really want to pull in the signals, you'll need to put up an outdoor antenna. Discones (an antenna which provides good reception from all directions) or ground plane antennas are the usual choice. Placed at rooftop level (at the peak of the roof if possible) these antennas allow your scanner to reach out beyond any obstructions that may affect an antenna at ground level. Try to get the antenna up as high as practical, attached to a strong pole and secured with guy wires. Beware of nearby power lines or trees. Mount the antenna as far away from these as possible.

Higher is better but that doesn't mean it's perfect. There is a disadvantage, too. When a signal is forced to travel, say, 200 feet down a coaxial cable to your scanner input, the trip saps the strength of the signal. Thus, you should try to keep your cable runs under 50 feet and use the best coaxial cable you can afford: RG-6 (available at most any radio or TV parts supplier) works nicely. Reception can be tricky, as variable as the weather sometimes. You'll fight it nearly every day you scan, but that's also part of the fun.

EQUIPMENT FOR MONITORING

The typical police scanner now covers much more that just law enforcement frequencies. The current crop of units are capable of scanning through thousands of different frequencies and storing your favorites in hundreds of memory channels. It is quite possible, even routine, to continuously monitor the frequencies of all local agencies in your area using a single scanner. The venerable Realistic (Radio Shack) PRO-2006, PRO-43, PRO-46, the AOR AR8000, Icom's IC-R1, and Bearcat's BC895XLT are just a few examples of modern scanners that take it all in stride.

There are many fine scanners on the market, but the first questions any potential police monitor seeking to buy a scanner must ask are: "Will I do most of my scanning at home or away? Or would I like to do both?" Next is the all-important question of: "Do I intend to listen to other services besides police?" And, you'll certainly want to ask yourself "How much do I want to pay for a scanner?" Scanner receivers range in price from around $80 to several thousand dollars, depending on features and manufacturer. Don't buy the cheapest - or the most expensive. Get something at a moderate price that will get you started and do the job. You can always "trade up" later.

When we think of scanners we most commonly think of base units that sit on a desktop at home. These are the most popular type but they don't lend themselves to mobile work - you can't take them to the zoo, the racetrack or the air show. Mobile scanners which mount in a vehicle are ideal if you do a great deal of driving, and handheld units that run on batteries and can be taken anywhere are other choices that may give you greater flexibility. Remember that the sensitivity isn't necessarily better on base units; handhelds are just as sensitive as their larger cousins, they just come in smaller packages.

Most serious scanner listeners tend to want one of each: a base unit for home use, a mobile for in the car, and a handheld. It's a matter of taste and what you can afford. Remember that a handheld can be easily adapted for use both at home and in the car. Think about flexibility when you go to make a purchase.

Before you sign your check or hand over your plastic make sure the scanner you're about to buy is capable of receiving 800 MHz frequencies. Though all scanners cover police frequency ranges, not all of them cover the 800 MHz range, which is where many police communications are moving. Your department may be next, if they haven't already upgraded their systems.

There are all kinds of gadgets and goodies available to supplement scanner monitoring. Here are my top six:

1. The most important is also the simplest. Always have a notebook and pen with you. You'll need them to keep track of active frequencies, new frequencies you may stumble across, codes, unit numbers or information you've been given by other listeners. Pocket size bound notebooks are best for portability and can be easily tucked away. I don't recommend notebooks with perforated pages for field use because they can tear free and spread valuable information into the wind. Three ring notebooks work great for home cataloging of frequencies and can be kept next to the scanner. Never be without pen and paper. Many police officers make it a point to carry two pens, a practice everyone should adopt. There's no telling when you might run out of ink just when you need to write down that elusive undercover task force frequency.

2. Frequency Counters - These devices will display the frequency or tune in near-field transmissions from just about any transmitter if you're close enough. The Optoelectronics Xplorer is an example of a near-field frequency counter. This company is in Ft. Lauderdale, Florida, and may be reached at (305) 771-2050.

3. Frequency Lists - No listener does much monitoring without a good list of frequencies nearby. There are many commercially available books - such as *Police Call* (Hollins Radio Data) and *Monitor America* (Scanner Master) - which list frequencies for areas across the country. The best list, however, is the one you compile yourself. Keep a list of what frequencies are already programmed into your scanner and another list of new frequencies you've found.

Computer database programs such as Grove Enterprises' CD Rom or the ScannerBase frequency guide for Windows also work great for keeping lists, and allow you to customize hard copy output. Try printing out a list of frequencies and cutting them down to 3x5" size to carry with you. Laminating them will keep them dry and clean. Whatever you use and however you do it, keep a good list of active frequencies, codes, call signs, names and numbers and you'll soon have a library of lists that will make your listening much more productive.

4. Antennas - This subject has taken up entire books. The technicalities are beyond our scope here, however there are a few tips that may prove valuable. Any scanner can benefit from an outdoor

antenna. At home use a discone or ground antenna.

For mobile use, mount a permanent or magnetic mount antenna on the roof or trunk. Handhelds are helped by extendable whip antennas to replace the common helical or "rubber duck" antenna. With all antennas, remember that bigger is not necessarily better, especially if you're trying to meet a landlord's requirements and need to keep a low profile.

5. External Speakers - A good external speaker will often outperform the speaker supplied with your scanner. This is true for both base and handheld units. Base unit speakers are often mounted on the top or bottom of scanners and can be blocked by tables, bookcases, or enclosures that will muffle the audio. If you have an external speaker plug on the scanner (most do) try hooking one up. This will work great in a car, too, where noise can become a real problem. Be sure to check the plug on the end of the speaker cable: some scanners accept only 1/8" mini-plugs, others 1/4" and some 3/32". You may need an adapter, which should be available at any good electronics supply store.

6. Earphones - These are great for private listening both at home and at noisy places such as air shows. Bear in mind that stereo headsets will require an adapter to allow you to hear audio from both ear pieces when plugged into your scanner. If you purchase monaural headset or earphones you won't have that problem. Remember that in most places it's illegal to drive with both ears covered by earphones. Now that you've been warned there's no reason why the next call you hear on your scanner will be from the patrol unit behind you, right?

HOW TO SET UP YOUR SCANNER FREQUENCY BANKS

The challenge of monitoring police communications is to set things up so that you can sit back and enjoy the action without worrying if you're missing something. The worst moment comes when an officer says he's switching to a tactical frequency and you hit the scan button only to find that he's disappeared because you didn't program that frequency into your scanner's memory bank. Each police department has its own method of arranging frequencies and each state has its own system of statewide law enforcement mutual aid channels. The key is to find out what system or approach is used in your area.

Once you have a good list of frequencies for your local department as well as county and state, devote an entire bank of your scanner's memory to these. This bank might hold 20 or 40 channels, depending on your scanner. If you don't have enough frequencies to fill up the entire bank, lock out the unused ones. Locking out unused frequencies speeds up scanning by bypassing the empty channels. If you have more frequencies than channels in a single bank continue programming frequencies into the next bank. Whatever it takes. I find it best to use the first bank for police frequencies. Channel 1 holds the Dispatch frequency for my area and subsequent channels hold other Dispatch frequencies. Once I have all the Dispatch frequencies in memory I move on to the Tactical frequencies, then Car-to-Car and Mutual Aid. Then I enter any emergency frequencies, detective's channels, county intersystem, statewide law enforcement channels, etc. Somewhere in there you might like to include the Inquiry or Information channel, but I find that this hangs up the scanner with routine warrant and license checks when I'm trying to listen to other activity. You might want to retain this frequency in the memory, but keep it locked out until it's wanted.

TWO SCANNER MONITORING

Many scanner hobbyists own two (or more) scanners, usually a base (desktop) unit for home use and a handheld for out in the field. At home listener's use the base unit and often ignore the handheld. That's a waste of scanning resources! Plug that handheld into an AC wall outlet and use it to keep an ear on police frequencies while you use the other scanner to hunt for new frequencies! You can even monitor two seperate agencies, one on each scanner. My location is close to the boundary of another department so I use one scanner to monitor my local agency and the other to keep abreast of what's going on across the city line.

Monitoring more than one scanner can certainly amount to an earful, but it's very helpful when you want to listen in on police and fire frequencies simultaneously without missing any of the action. At first it may seem overwhelming, but gradually you'll learn the knack of keeping track of transmissions on both scanners. Back in the days of crystal-controlled scanners, former KFWB-Los Angeles Program Director Jim Hawthorne kept *five* scanners going to keep track of as many different frequencies. So using two scanners isn't really so tough!

4

FINDING LAW ENFORCEMENT FREQUENCIES

Before you can monitor anything you need to know where to listen. Scanners can cover thousands of frequencies in a myriad of bands: Police, Fire, Emergency Medical Service (EMS), Aircraft, Business, Forestry, Motor Carriers, Maritime, Special Industry, Federal, etc. Where and how do you find the channels used by officers in your hometown, patrolling your neighborhood? They have to be somewhere, but among all the possibilities it can make for a long, even frustrating search. Let's try to narrow it down.

RESOURCES FOR FREQUENCIES

Let's start at the beginning, as if you have just purchased a scanner and are sitting down to listen to police for the first time. Step number one is to find out the exact name of the agency responsible for law enforcement in your area. Does your city have its own police department? Or is your town protected under a contract with another police or sheriff's department? Look in the local government section of your telephone book for answers.

Books

The next step is to get your hands on a scanner book which lists public safety frequencies, which encompasses police, fire, and emergency medical services. We're in luck here because there's no lack of information on these kind of frequencies. Best known are the *Police Call Radio Guides* published by Hollins Radio Data and available at Radio Shack stores. These books list frequencies directly from FCC licenses as they are issued. The listings are in two forms, alphabetically by state and also by frequency. There are currently nine such volumes, each covering different portions of the United States. Make sure you have a current copy of the edition which covers your area. because frequencies may change over time.

Other frequency books that will be of help are *Monitor America* published by Scanner Master. Also the *Master Frequency File* (federal

law enforcement) by James Tunnell and Robert Kelty, (published by Artsci) and Robert Coburn's *Official Guides* for various states. Visit your nearest radio communications store or take a look at a current issue of one of the radio monitoring hobby magazines for these and other useful books.

Magazines

Speaking of magazines, those dedicated to the radio listening hobby can be a great source for frequencies and other information. Many feature columns where frequencies are exchanged and new or changed frequencies are reported. Try *Monitoring Times, Popular Communications* or *National Communications Magazine.* You might consider placing a classified ad in one of them, requesting contact with other listeners in your area. Another good reason to read these magazines are the many advertisements for scanners, scanner accessories, books, computer software and other items designed to increase your listening prowess.

Clubs

Scanner clubs are one of the best bets for finding like-minded people who are willing to share their knowledge and frequencies. Many organizations operate on the strength of member input alone. Clubs such as the Radio Monitors of Maryland, Scanning Wisconsin, All Ohio Scanner Club and the Palm Beach County Scanner Group and any number of others trade information and frequencies. There is an excellent listing of clubs (including contact information) on the Grove Enterprises Web page at www.grove.net. Club lists and information is also available at the Web site of the Association of North American Radio Clubs: www.anarc.org.

Other Monitors

The best way to find out about local police frequencies is through contact with other listeners in your area. Local clubs (even amateur radio clubs) can be a gold mine of information. Check the newspaper's community section for meeting schedules or ask at an electronics store. You may even want to consider starting a club yourself.

Radio Retail Outlets

Before you walked out of that electronics store where you bought your scanner did you ask the clerk about local frequencies? It's not

uncommon to discover that these folks listen to scanners themselves. Maybe they can help get your started, or at least point you in the right direction. It doesn't hurt to ask.

The Internet

If you're on line you already have access to one of the best sources of radio related information: the Internet. In addition to the "real world" based clubs, there are many Internet lists that cater to scanner enthusiasts: Scan-L, Milcom, Scan-DC, Trunkcom and others. By hooking up with a scanner club or list you'll have access to a world of scanning information that will greatly increase your knowledge and listening capabilities.

Point your browser to one of the search engines such as Yahoo (www.yahoo.com) or AltaVista (www.altavista.digital.com) and search under the keywords "scanners," "radio" or "frequencies." New sites pop up every day so check often and follow the links. You never know what you'll find. Don't neglect the Usenet newsgroups where discussions on scanning can be found in places such as alt.radio.scanner. Also, look for radio-related forums on the major online services such as CompuServe, America Online, and Prodigy.

Police Departments

Yes, you can show up at the front desk of your local police department and ask what frequencies they're using, but I don't recommend it! This approach should be a used *only as a last resort*, unless you happen to know a police officer personally, or at least well enough to ask for this kind of information. Frequencies are available from a wide variety of sources so try this approach only if you've tried the other avenues several times but have been unsuccessful in all your other attempts.

FINDING FREQUENCIES WITH YOUR SCANNER

Of all the places to look for police frequency information the best location is right under your nose, or rather, right under your fingertips. All of the major scanner models have a search function, which allows you to program in the lower and upper limits of a desired search range and set the unit to automatically look for active frequencies. To prevent missing transmissions, limit your search to short chunks of the spectrum at a time, such as 154-155 MHz, then 155 to 156, and so on. The frequency ranges used by police communications ranges are listed in the appendix.

Don't give up if you don't come across anything after a few minutes; this kind of search takes time. Give it a few weeks of continuously scanning the same ranges and see what you come up with. Keep a record of the "hits" or active frequencies you find and return to check these again until you can determine what agency is using them. It will take time but you will eventually come across the frequencies you're looking for. Doing your own sleuthing is fun, and you'll enjoy the satisfaction of having dug these out yourself.

COUNTERS

Another way of discovering police frequencies is by using a frequency counter. All you need to do is get close enough to a police officer who is using his radio and then just read the frequency off the frequency counter display. This method requires you to do some legwork, and many scanner owners are understandably hesitant about following cops around, flashing scanners and frequency counters. But if you're *very* discreet and you stay well out of the officer's way you should have no problems.

Frequency counters require you to get as close as possible in order to obtain a good, clear reading. These gadgets are wide coverage devices that are easily confused by other transmissions. As a rule of thumb, if you're within a few car lengths of the transmitter you have a good chance of snagging the frequency. The best way to get close is in your vehicle. However, it's not a good idea to try and drive your vehicle while at the same time attempting to capture the frequency of a patrol car ahead of you. It's smarter to ask a friend to do the driving, leaving you free to devote your attention to monitoring the equipment You can also park near a patrol car or an officer and switch on your frequency counter. (See the "low profile" monitoring tips we'll discuss later.) Be sure to keep the device out of sight. Most importantly, *never impede the flow of emergency traffic, never cross police lines or cones, and never get in a officer's way.* Getting the frequency isn't that important; safety is!

Try parking in or near the police department parking lot and using your counter to pick up frequencies in use. Many departments have their communications center in the same building as their headquarters. If you show up around shift change, officers testing their radios will give you an excellent chance of success. If you're close enough, use a short helical antenna - a "rubber duck" - on your counter to minimize the interference and false readings caused by other transmissions.

5

TECHNICALITIES OF POLICE RADIO SYSTEMS

There are a few technicalities that you should know in order to successfully monitor police radio. This information will serve you well, not only for this kind of listening, but also for others types such as fire and emergency medical.

REPEATERS AND SIMPLEX

You will often see the letter "R" after the numbers in frequency directories. This stands for "Repeater." A repeater, in its least technical sense, is a radio that receives on one frequency and retransmits (repeats) the signal on another frequency. The frequency used to receive communications is called the Input Frequency and the channel used to rebroadcast is known as the Output Frequency. Repeaters have greater power and are usually located on hilltops or high buildings so they can be heard over greater distances than if the call went out using only the strength of a patrol car's radio. What a repeater means to the listener is the ability to monitor calls more clearly and easily. Mobile units that might be out of range of your antenna due to terrain or distance can be heard loud and clear through the repeater.

Simplex systems don't use a repeater. These communications are line-of-sight, meaning that if your receiver's antenna is obstructed by terrain, trees, or tall buildings, you probably won't be able to hear simplex communications. Simplex is often used for Car-to-Car and Tactical communications where all units are near to each other (line-of-sight) and the services of a repeater are not needed. This also makes it more difficult for unauthorized persons to eavesdrop on the communications. Detective and undercover operations use this feature quite often. Your antenna makes all the difference; a good outdoor antenna will bring in transmissions you wouldn't hear otherwise.

In frequency listings it's usually best to choose the frequency marked "R" over one marked "S." However, don't totally disregard simplex frequencies because you may hear communications here that circumvent the repeater.

VEHICULAR EXTENDERS

Extenders are simply vehicle-mounted repeater systems. These are used in some areas to extend the range of portable radios by repeating the transmission through the patrol unit's own extender radio and antenna. Vehicular extenders are great for officers in remote back country areas or places of bad reception.

TALKAROUND

Talkaround is very much like simplex; direct communications without using a repeater system. Talkaround usually occurs on the repeater output frequency. When one officer asks another to "go to talkaround" you may lose the conversation if you are not in the line-of-sight of the transmitter. Run through the police frequencies in your scanner banks and count yourself lucky if you come across the same two officers on talkaround. Officers use this to lessen their chances of being overheard. Nine times out of ten, however, it doesn't do them any good.

HIGH TECH POLICE RADIO

In 1992, when *Cop Talk* was written, monitoring law enforcement radio was still a fairly simple business. Most systems were analog rather than digital and transmissions were on specific frequencies dedicated to specific purposes. Today, however, technology has smacked the listener in the face (or the ear) with new types of high tech radio gear. In some areas it isn't as easy to listen in anymore. 800 MHz trunked and digital radio systems are putting roadblocks in the way of this favorite pastime. Fortunately, all is not lost. The new systems can still be monitored.

800 MHZ TRUNKED SYSTEMS

The latest craze in police radio are 800 MegaHertz trunked radio systems. These million dollar systems utilize five or more frequencies in the 800 MegaHertz band, which rotate so that when a transmission is made a computerized controller automatically routes the call to a free channel and switches all other radios in that talk group to that channel. For listeners with standard scanners, the transmissions seem to hop around. Every time a unit unkeys its microphone the computer listens for a reply before it releases the frequency and instantly assigns a new one for the reply.

It is quite possible to monitor these radio transmissions by programming your scanner with each of the 800 MHz frequencies assigned

to your local police agency. One of the frequencies will be known as the Control Channel, used by the computer to send information to mobile unit radios. You'll know it when you hear it: it emits a loud, roaring buzz. Use your channel lock-out feature to lock this one out (see your scanner's operation manual for instructions). Be sure to check the locked out channel each day, however, because the control channel can rotate among assigned frequencies. If it changes all you need to do is unlock the old frequency and lock out the new one. If a subsystem is in operation, you'll find more than one control channel in use. Listen to call signs and users to determine who uses which frequencies.

One of the pitfalls of listening to a trunked police system on a standard scanner is that the system might also carry the communications of fire, rescue and city public works, as well as those of any number of other users. Each will have their own talk group on the system, so while the same frequencies are used, each user's communications are "invisible" to the other users. A conventional scanner is not set up to discriminate between the transmissions. You'll need a quick hand in order to follow the communications of a particular agency.

Let's say that the police dispatcher calls a patrol unit, then unkeys the radio to wait for a reply. Your scanner continues scanning the other frequencies in the system and stops on one where a city crew is talking about a water main break. Hit the SCAN button immediately and keep hitting it as necessary, to bypass other conversations until you find the police call again. As soon as the conversation ends on one frequency and the carrier or open frequency is dropped, communications will move to another frequency. Avoid using your DELAY feature, as it will just slow things down and cause you to miss calls.

Some trunked systems, such as that made by GE, utilize a series of tones after the microphone is unkeyed. These tones will "hang up" your scanner. Others systems have tones that rotate through the voice frequencies. You'll need to bypass these manually.

Of course, all this hassle and hubbub may be done away with if you purchase a TrunkTracker scanner such as the ones now being sold by the Uniden company. Unveiled in January, 1997, the TrunkTrackers are wildly popular with listeners, who appreciate the ability to monitor trunked systems as they should be monitored. Unfortunately, at this writing, TrunkTrackers can only follow 800 MHz transmissions on Motorola analog systems. If your local department is using a digital system or equipment other than Motorola, the TrunkTracker will be of no use.

With the TrunkTracker, the listener programs in all of the frequencies of the system and hits the SEARCH key. The scanner takes over, hunting through the frequencies until it finds the control or data channel, from which it will take its information. Immediately after that it will begin to follow talk groups, which are basically the old "channels" that we once knew and loved. If you want to monitor one talk group, say the one belonging to Police Tactical Operations, you can instruct the TrunkTracker to monitor that particular talk group. Just as hunting for frequencies was half the fun of conventional scanning, figuring out which talk group belongs to which agency or division is half the fun of trunked monitoring. TrunkTrackers can monitor conventional frequencies, too, but so far they can't follow a trunked system and monitor conventional channels simultaneously.

Due to the high costs of trunked systems it's unlikely most smaller departments will install these anytime soon, unless they join a countywide or a larger city system, thereby sharing in the cost.

COPING WITH ENCRYPTION

It isn't uncommon to find a police department "scrambling" or voice-encrypting communications related to undercover details and conversations that, for one reason or another, they would rather not be overheard by the general public.

The sad fact is that criminals sometimes own scanners and use them to try and stay one step ahead of law enforcement. This makes it difficult, to say the least, for officers to do their jobs and maintain their personal safety on the street. Encrypting communications is one of the ways in which criminals with scanners can be defeated.

Many agencies find communications encryption to be cost prohibitive, but officers tell me that the threat from criminals listening to their communications is very real. More than one sensitive operation has been blown because the bad guys were able to listen in and keep track of police efforts to apprehend them.

Members of the news media also pose a problem for police operations, as most news desks and reporters routinely use scanners to alert them to breaking news. In one case a reporter overheard an officer relay the telephone number of a residence where a hostage situation was occurring. The reporter dialed the number and attempted to obtain an over-the-phone interview with the criminal. As you can imagine,

encryption can and does eliminate many of the problems with security that officers on sensitive operations encounter when they use their radios.

A scrambled channel may sound like unsquelchable static or high-pitched gibberish, depending on the encryption method used. There are several methods of scrambling communications, ranging from simple speech inversion to the Digital Encryption System (DES), popular with federal law enforcement agencies. Some departments, such as the Provincial Police Metro Divisions in Ontario, Canada, use a constant noise called "cloaking" to mask radio transmissions.

At one time, you could buy and build a kit that would defeat speech inversion, but doing this today is illegal, according to the Electronic Communications Privacy Act of 1986. If your local department scrambles some of their communications you could investigate unscrambling methods, but I don't recommend this. If you're like me you realize that some police communications just aren't meant for everyone's ears. For officer safety and the success of the operation we must accept the fact that voice encryption is sometimes necessary. Listening to normal unscrambled channels will often clue you in to what is happening.

COMPUTER AIDED DISPATCH AND MOBILE DATA TERMINALS

The days of dispatching via voice radio are slowly changing in many areas as new computer controlled systems are introduced into the police radio service. It is unlikely that voice communications will ever be totally replaced, but the computer technology in police communications is definitely on the increase.

Computer Aided Dispatching (CAD) eases the dispatcher's workload considerably by placing information on computer screens, with each screen able to display different information, such as unit status and incident logs and records. In addition, field units are able to reply to dispatcher messages by pressing buttons on dashboard-mounted control units which send condensed message bursts to the dispatcher's computer screen, allowing quick updating of status screens.

Mobile Data Terminals (MDT) are vehicle-mounted control units which talk to the CAD system electronically. An MDT looks much like a laptop computer, complete with screen and keyboard. Messaging between dispatcher and field unit and between officers in the field can be done without the use of voice.

After the infamous Rodney King/Los Angeles Police Department police brutality incident LAPD officers used their MDT's to chat about what had happened. Little did they know their MDT communications were being monitored and recorded by headquarters. The MDT exchanges were later released to the media.

CAD systems save dispatchers a great deal of time and effort by taking some of the work out of the job. A CAD system functions under a basic four-step process.

1. A police department receives a telephone call, counter appearance, or other communication requesting assistance from a citizen.

2. The police 911 call taker or dispatcher enters the basic information into a computer which matches the call location with a geographic base file and routes it to the appropriate dispatcher.

3. The computer also assesses the availability of officers on the appropriate beat and assigns one of them to answer the call. The system might also be programmed to search local, state, or national databases to determine if there are any records for the persons involved in the call. Information about prior calls for service from that address (premise history) can also be displayed.

4. Then the computer sends the bulk of this information, in condensed form, to the dispatcher responsible for that geographic area and from there it is dispatched by radio or MDT to a field patrol unit.

While computer dispatching assists and eases the load of dispatchers and the officers in their day-to-day operations, it presents a unique problem to the police monitor; instead of normal voices we hear only beeps, blats and growls. There is computer software available via shareware that supposedly can be used to decode some MDT transmissions. Unfortunately, however, the law says the use of such aids is illegal.

"The only objection that I have to scanners is when burglars and armed robbers use them to monitor police calls. Otherwise, the public can be very helpful when they hear something has happened in their neighborhood and keep their eyes open for suspects, etc. before the cruisers get there."

- Capt. Michael Hanna
Flint, MI Police Department

6

RADIO JARGON

There is no facet of police communications more befuddling to a listener than radio jargon. Codes and numbers are used in law enforcement to keep communications brief yet understandable. Officers are often not in situations conducive to long-winded dissertations. A few words are all time might allow, and by using codes, a few words are all that are needed.

Unfortunately, what makes for quick and clear radio conversations for police officers makes for confusion on the part of the police listener. Civilians are hard pressed to enter the world of law enforcement without actually pinning on the badge. Therefore, understanding radio communications means understanding police work. It takes time and research, but most of the codes you hear can be puzzled out, and some are even self-explanatory.

You should start by using all of the resources for finding frequencies mentioned earlier. These same sources can often provide you with radio codes. A police officer is also an excellent person to ask, if you know him well enough. But please don't accost officers on the street seeking information. They're busy enough.

Codes and abbreviations can differ widely between departments, cities, and states. There are no guarantees that the California code for a stolen vehicle (10851) will be the same in New Jersey or Iowa or Texas. Check the appendix for a sampling of various codes used in different areas of the country.

POLICE PHONETIC ALPHABET

With some exceptions the phonetic alphabet of law enforcement is the same or similar throughout the country. Each letter of the alphabet is given a name, such as "Adam" for the letter "A," "Boy" or "Baker" for the letter "B" and so on. This makes for a minimum of confusion when reading suspect names, addresses, or license plates over the radio. Even with perfect reception a "V" can be heard as a "B," X's sound like S's, and in all the confusion the bad guys could be getting away. By using

the phonetic alphabet an officer is assured that everyone understands what he's talking about.

A	Adam
B	Boy
C	Charles
D	David
E	Edward
F	Frank
G	George
H	Henry
I	Ida
J	John
K	King
L	Lincoln
M	Mary
N	Nora
O	Ocean
P	Paul
Q	Queen
R	Robert
S	Sam
T	Tom
U	Unit
V	Victor
W	William
X	X-ray
Y	Yellow
Z	Zebra

POLICE INCIDENT AND DISPOSITION CODES

Codes are part and parcel of a police officer's job. They describe everything from transmission readability to fires, murders, and requests for registration checks. Just about any situation or condition you can think of has a designated code. Remembering them all takes time and practice.

Each police agency has its own set of codes. There has been a trend to consolidate codes by region (city, county, and state) and indeed, you will find some commonality. However, not all departments heed this so

you'll find a wide variety of numbers used. In general, most departments follow the "10 codes" officially suggested for use by the Association of Public Safety Communications Officers (see the appendix for a list of APCO codes). I guarantee, however, that every department has expanded on these codes. In fact, a whole new series, called "11 codes," is now being used in many areas.

Codes are commonly divided into two major categories: Incident Codes and Disposition Codes. As you undoubtedly guessed, Incident Codes serve to define the type of incident a police unit is responding to or is handling. An 11-80 might mean serious injury accident, while 10-78 might mean a fire. Disposition Codes define the outcome of a radio call. Signal 18 might mean the call was unfounded; 10-63 might mean an arrest was made.

In law enforcement communications, units dispatched to calls for service are always advised of the type of call (if known) and units usually advise the dispatcher of the outcome of the call as they clear back into service. Codes are used heavily for this and knowing what they mean makes the difference between understanding what's happening out there on the street and being totally in the dark.

Deciphering codes can be a long uphill battle if waged alone. Your best bet is to find a good frequency guide for your area, as mentioned earlier. These usually carry a code list. Also, go searching for some scanner listeners in your area. It's a good bet they will have the codes already, which will save you time and frustration. Remember, though, that half the fun of listening is figuring out the terminology yourself.

COP SHORTHAND: ABBREVIATIONS IN COMMUNICATIONS

"9981, 10-8 RF, AF. I'll be *en route* RDF with one. Can you give me CAD and TOC?"

Huh? That's cop talk at its most cryptic; a statement so filled with pseudo-gibberish that you're probably tossing up your hands in incomprehension. In addition to 10 and 11 codes, officers use a specialized lingo to communicate. These are often just abbreviations and with a little listening you can figure out what they mean.

Try the above statement: 9981 is the patrol unit number, which lets the dispatcher know who's talking. 10-8 is a standard 10 code for "in service." RF means "Report Filed." AF is "Arrest Felony." These are common disposition codes, but trying to decipher the rest of the statement is where it gets tricky. Obviously, this officer is *en route* to a

place called RDF. We know he's made an arrest ("*en route* with one"), so it's pretty clear that he's taking his prisoner to jail. A check of the phone book for the names of area jails reveals the River Detention Facility…RDF! Next, the officer asks for CAD and TOC. Let's listen to the dispatcher's reply:

"10-4, 9981. CAD is 379, TOC was 1815 hours."

As explained earlier, CAD is short for Computer Aided Dispatch, the computer system dispatchers use. CAD assigns a number to each call for service and that's what the officer is asking for. Instead of saying "you know that call you sent me on with the guy who was drunk and carrying a knife? The one at about 6 p.m. on Westlake Drive?" he just says "in reference to CAD 379."

How about TOC? From the dispatcher's reply we can guess that he's giving the officer a time. Therefore, TOC stands for Time of Call.

That wasn't so hard, was it? In police work, just about everything is abbreviated. Not only does it make for easier comprehension, it helps to keep radio traffic to a minimum (and listeners in the dark!). Not everyone who listens in is a solid, upstanding citizen. I'd much rather hear an officer say "meet me at M and M" than "Meet me at the corner of Mission and MacKenzie streets." If you listen long enough, though, sometimes officers will use the long version and you'll be able to add another abbreviation to your list (you are writing these down as you hear them, aren't you?). Even if you don't understand the code or abbreviation, make a note of it anyway. Something you hear later on may make it clear.

Another tip about codes: put in some study time at the public library. Check out the volumes that deal with your state's penal code and city or county laws and regulations. You will sometimes find that the codes you hear on the radio are taken directly from these statutes. For instance, in California, when officers talk about a burglary, they use the penal code 459.

The key to figuring out codes is to listen! Listen to the radio exchanges and consider the context within which the code or abbreviation is used. This often gives the meaning away. Sometimes you get lucky and an officer or dispatcher uses the plain language version. Write it down!

UNIT NUMBERS: HOW TO IDENTIFY WHO'S WHO

Most of us remember the television show "Adam 12," about a pair of Los Angeles Police Department patrol officers. Their unit number reflected the actual numbering system still used by the LAPD today. "One" (used in radio calls in the TV show) indicates the area - Central - and "Adam" indicates the type of patrol unit, in this case a two-officer patrol. "Twelve" specifies the individual unit- the number twelve patrol unit in the Central division.

Unit numbering for law enforcement agencies varies, with each department having their own methods to identify their radio units. Some use the One Adam 12 method, some the three digit method (such as "244") and still others use variations on those themes: "David Three," "Eight Boy," "21 Sam," etc.

Unit numbering serves three purposes: it identifies the shift or unit location, the type of unit, and the individual unit number. There are infinite variations. Let's take another example:

A city police department in California uses the three digit method for unit ID's. The first number indicates whether the unit is a morning, day, or evening shift - 1, 2, or 3. The second and third digits tell the type of unit - 11 to 29 are detective units, 30-39 are traffic units, 40-65 are patrol units. Occasionally a letter is used after the three digits to further define the unit: Mary for motorcycles, Tom for traffic units, King for K-9, Sam for Sergeants.

If you can't obtain a list of the unit numbering system for your local department do the next best thing: figure it out yourself! Write down every unit number you hear until you have a good size list. Listen to the type of calls units are dispatched to. Maybe "One Tom Seven" is responding to an awful lot of traffic accidents and road hazards and doesn't seem to get calls for 911's or ringing alarms. That indicates he's probably a traffic division unit. What about "311 David"? You've been hearing him in the evening, but infrequently and usually in connection with a crime already committed. Maybe he's a detective unit.

Be extra alert when you hear an officer ask the dispatcher to send a traffic unit, patrol unit, investigator, etc. to his location. Listen to who the dispatcher sends and note the ID number. This is the quickest way to determine who's who.

Pay attention to any police vehicles you see in your travels around town. Some may have their call signs printed on the fender, hood, or roof. Others have only vehicle ID numbers here, rather than call signs, so look carefully.

The following are some of the designations used to define radio units. Use them as a general guide, bearing in mind that different departments may use the same lettering with different definitions.

    CHARLES: Captain
    HENRY: Homicide
    JOHN: Juvenile
    LINCOLN: Lieutenant
    SAM: Sergeant
    IDA: Investigator
    MARY: Motorcycle
    TOM: Traffic
    KING: K-9
    PAUL: Patrol
    VICTOR: Vice

RADIO UNITS

Once you begin to listen to police radio you will confronted with a bewildering array of call signs. Everyone using the radio is given a set of numbers to use as an identifier. This is the smoothest way to run a communications department and at the same time avoid costly confusion. But, who are these people behind the numbers? What type of units are they and what are their jobs?

Uniformed Division

The most visible and well-known part of any law enforcement agency is the uniformed division, which is composed of the officers you see on the streets every day. The uniformed division includes:

Patrol Officers who respond to the bulk of the radio calls which require the intervention of a peace officer. The senior officer on patrol during a shift is traditionally a sergeant, who is called the Patrol Supervisor.

Traffic Officers who handle injury and non-injury vehicle accidents, issue traffic citations for a variety of moving violations, provide traffic control, conduct driving under the influence evaluations, perform vehicle safety and compliance inspections, and investigate and make arrests.

Community Service Officers who are usually non-sworn department personnel who handle routine calls not requiring sworn or patrol

personnel. CSO's in many departments serve as crime prevention and awareness officers and free sworn or uniformed officers for more critical calls.

Community Oriented Policing Officers who are the modern incarnation of the beat cop, patrolling unassigned sectors of the city and working closely with residents and businesses to prevent crime and improve safety.

Investigations Division

When the flashing lights fade and the scene is under control, the investigators move into action.

Detectives investigate and make arrests for crimes such as homicide, theft, assault, rape, burglary, robbery, sexual abuse, etc. They also perform stake-outs and surveillance.

Crime Scene Technicians are vital to criminal investigations and often make the difference in solving or proving a crime. These men and women traditionally respond to assist investigators by photographing crime scenes and collecting evidence. They also transport evidence to the laboratory for analysis by Forensic Science Technicians.

Special Units

A law enforcement agency must assign special units to ongoing investigations or special operations requiring uniquely trained officers.

K-9 Officers work with specially trained dogs on uniformed patrol and special operations to apprehend criminals, search vehicles and buildings, sniff out narcotics or explosives, and find lost or deceased individuals. Some K-9's are cross-trained to perform more than one of these jobs, others are specially trained for one activity, such as searching for narcotics.

Bomb Squad Officers handle explosive and hazardous devices by containing, removing, and rendering them safe.

Narcotics Division Officers are responsible for investigating and making arrests for cases of illegal narcotics use, manufacture, or sale.

Gang Unit Officers keep track of gang members and investigate cases of gang violence.

Fraud Unit Officers investigate deceptive currency, documents, and practices.

Fugitive Apprehension Unit Officers work with local, state, and federal services to locate and apprehend fugitives from justice.

Air Support Unit Officers serve as pilots and observers aboard police department helicopters and other aircraft, performing search and rescue and medical evacuation, providing air support to ground officers, and transporting special unit officers.

Marine Unit Officers operate aquatic vehicles for harbor, lake, and waterway patrol and enforcement. Some of these officers also serve as search, rescue, or recovery scuba divers.

Special units are a part of most medium to large size police departments. Your local department may have all or only a few of the units named above. Regardless, by careful listening, you can determine which are active in your area.

KEEPING TRACK BY KEEPING NOTES

A law enforcement listener's most useful tools - next to a scanner - are pen and paper. Write down numbers, codes, abbreviations, frequencies - anything and everything you hear on your scanner. Collect lists of frequencies in use, with notes about what type of transmissions you hear on them, and any number or code used to identify that frequency. Write down unit numbers and types of calls assigned to those units. All information is important when compiling a profile of a police department's radio communications. Don't disregard anything, no matter how trivial it sounds. Using these methods, I have been able to identify units, their usage, beats, locations, and even the names of individual officers. I have learned voices and been able to associate them with units even before an identification was given over the air.

Purchase a good street map of your area and keep it nearby as you listen, You might want to have the map laminated so you can mark it up as you follow a police event. That way you can wipe it clean and use it over and over. Glue the map to some sturdy backing or put it on a bulletin board near your scanner. Note the locations of calls and what units are sent in response. If it takes the unit a short time to arrive at a call, then he most likely was already in the area and you can assume he handles that beat regularly. If he takes longer he probably had to respond from farther away. Make note of that, too. Soon you'll begin to get a pretty good idea of what units cover specific areas of the city.

Dispatchers often refer to "beats" when dispatching calls. "5 Ocean, on the 54 Beat, at 1165 River Drive..." Pencil in "54" on the map in the area of River Drive. Once you find other streets on the same beat you'll begin to define the boundaries of that beat. Do this with each beat in the city and before long you will understand how the beat system works in your department. Don't give up! It takes time and tenacity.

7

MONITORING LAW ENFORCEMENT COMMUNICATIONS

Let's talk specifics about police radio exchanges. What are you going to hear? What are the different frequencies used for? Where do you find undercover operations? Bear in mind, again, that your local department's communications system setup may vary. Even so, all communications are based on a standard format.

DISPATCH

Every department, no matter what its size, uses a central frequency or frequencies for dispatching units to calls for service. Dispatch might consist of one frequency or several, depending on the size of the geographic area covered by the department. Large departments such as Dallas, Los Angeles, or New York City may use a series of frequencies for dispatching, one each for the eastern, western, southern, and northern areas. In certain cases these might be broken down even further and other frequencies used for even smaller areas or zones. Departments which use trunked 800 MHz systems still use Dispatch frequencies, only these are broken up into talk groups.

Dispatch's job is to send out units to calls received at the station or the 911 Public Safety Answering Point, keep track of what units are available and where those units already on calls are located. Things can get hectic on the Dispatch frequency during peak hours, weekends, holidays, and (so the myth goes) full moons. The frequency may virtually shut down for emergency situations, such as when a unit is on a high-risk vehicle stop, a burglary in progress call, or a shooting. Officers may request what is variously known as "Emergency Traffic," "ET," "Clear the Air," "I need the air," or simply "The Air"- when they are involved in a dangerous situation and their communications must get through to the dispatcher. Many systems have an automatic beep tone which periodically warns other officers not to transmit unless they have urgent traffic.

Dispatch is the hub of all patrol communications, the place where dispatchers at the communications center keep track of all officers in the field. Needless to say, the Dispatch frequency should occupy center stage on your scanner.

INQUIRY

Also known as Information, Data, Wants/Warrants, or even Miscellaneous, this frequency is used for requesting license and registration checks, searches for wants and warrants information, checks on serial numbers of stolen property, missing persons information, and general information that would tie up a Dispatch frequency unnecessarily. Keep an ear on this frequency to hear all the details when an officer runs a person's license or registration or checks a plate number through the computer for hits. This is the only type of traffic you will usually hear on this frequency.

Inquiry has seen some phasing out in departments where Mobile Data Terminals enable officers to access state records information (often called "local") and the Federal Bureau of Investigation's National Crime Information Center (NCIC) through their vehicle-mounted computer systems. This type of direct access cuts out the middleman - the Inquiry dispatcher - speeds up records searches, and frees valuable air time. For a listener, this means that some routine information normally available over the air is no longer heard.

TACTICAL

Law enforcement agencies commonly designate one or more frequencies to serve primarily as a Tactical channel, a place away from regular Dispatch and Inquiry frequencies where officers engaged in special operations can communicate. Special ops can involve anything from routine surveillance to warrant services, searches, narcotics raids, SWAT missions, hostage situations, and more. These reserved channels may have one of many names. One of the following may be used in your area:

Tactical or "Tac"
Car-to-Car or "Cars"
Unit-to-Unit
Portables
Mobiles
Alert Net

> Special Unit Channel
> Ops
> Special Operations
> SWAT
> Field Communications
> Surveillance
> Operations Primary/Secondary
> Investigations
> Special Incidents
> Detectives
> Secondary
> Private

Each special unit within the department may have its own frequency which might bear the unit's name: Street Crime, Gang Suppression, Bike Patrol and the like. If it isn't used as a Dispatch, Inquiry, or Administration Channel keep an eye on it, as Tactical communications might be transmitted here.

In addition to dedicated department frequencies, some areas use county or state appointed channels for Tactical operations, especially if the mission is conducted in collaboration with another agency. Look for activity on Intercity, Interagency, Interservices, Mutual Aid, Citywide, or Intersystem channels. 155.475, for instance, is used throughout the United States as a VHF mutual aid channel for law enforcement. It's called by various names, including "Hotline," but it is most commonly known as NLEMARS ("nailmars") the National Law Enforcement Mutual Aid Radio System. On the 800 MHz band, the following channels are used for mutual aid:

> 866.0125
> 866.5125
> 867.0125
> 867.5125
> 868.0125

Tactical channel users include special units like SWAT (Special Weapons and Tactics), detectives, special divisions within the department such as Major Crimes or Auto Theft, and field officers. Practically anything goes on Tac channels and you may be surprised at what you hear. It's not uncommon to hear dispatchers talking to patrol units

about sensitive information or officers talking to each other about everything from the fascinating to the mundane.

Most Tactical communications on dedicated channels are broadcast without the use of repeaters so as to minimize the transmission range. The threat of being overheard constantly dogs police officers and disaster can strike if the wrong ears hear certain Tactical broadcasts. For this reason, most Tac transmissions use the Simplex system, which covers a shorter distance and doesn't use the higher power a repeater provides. Since these type of transmissions are short range they reduce the risk of being overheard.

CAR-TO-CAR

When an officer says that he is "switching to Cars" it doesn't mean that he's tired of his motorcycle or foot patrol assignment, but that he's moving from his current frequency to the Car-to-Car channel.

"Cars" is known by different names, depending upon the department. It may even be known as Tactical, but the use is the same: it's the frequency officers use to talk to each other about cases and suspects, exchange information, arrange meetings, and coordinate responses. The content on this frequency is typically very candid; it's the best place to hear the unvarnished truth about calls, cases, and situations. Information exchange between police officers can be vital and Cars provides the radio link.

Cars is usually a simplex frequency on conventional radio systems and therefore transmissions don't carry very far. You'll have to be close to hear the exchanges.

EMERGENCY

The name says it all. Emergency is a separate frequency usually set up by departments to handle emergency situations. This can range from search and rescue, SWAT call-outs, hostage situations, or homicides. If an 800 MHz system is in use, a signal is sent to the dispatcher on the Emergency frequency when an officer pushes the appropriate button on his radio. Either way, the only communications that take place here are of an urgent nature. Keep this frequency in your scanner's memory and check it often …you never know.

OTHER SPECIALIZED FREQUENCIES

Every police radio system is set up differently and your local police may have other frequencies assigned for units such as Detectives, Command, Special Operations, Custody/Jail, Court Service, Special Units, Paging, Air Unit, Juvenile Division, Burglary, Auto Theft, SWAT, and Simplex operations. It takes a lot of listening over a period of time before you can spot them and begin to sort them out.

There is a trend to provide all police agencies within a state with the ability to communicate with each other on a specific frequency. Such an ability is imperative in the case of a pursuit crossing city lines, for coordinating interagency assists and operations, information requests, etc. The frequency is usually known as Mutual Aid or Interagency. It may also be identified by a color code or acronym, such as CLEMARS, the California Law Enforcement Mutual Aid Radio System (154.920 MHz.)

It's an excellent idea to seek out these Mutual Aid frequencies and program them into your scanner's memory. Many departments run joint operations with other city, state, and federal agencies. Mutual Aid is where you may find them.

FREQUENCY NUMBERING

Every frequency used by a police department carries a designation to enable quick and easy identification. A dispatcher could ask a unit to switch to "154.235" but it's a lot easier to say "switch to Frequency Three." Not only does this provide a certain amount of privacy, it also ties into the way in which most police radios are set up. Pressing the switch marked "3" will put an officer on Frequency Three. Of course, with 800 MHz systems, individual frequencies don't carry designations so much as specific talk groups do.

Many departments use the F-numbering system. F-1 is Frequency One, F-2 is Frequency Two and so on. The F is commonly dropped in verbal communications. Also popular are color-coded frequencies such as Blue, Green, or Red channels. Some frequencies carry the designation of the unit or area that uses it: Narcotics, Barracks A, Traffic, Downtown, Harbor. Talk groups for 800 MHz carry ID's such as 564 or 20887.

Frequency directories often list the frequency designations alongside the numbers. If you can't discover them this way, listen carefully to the radio exchanges and write down the designations you hear. Gradually, through a combination of officer/dispatcher exchanges, you will be able to match numbers to frequencies.

CELLULAR TELEPHONES

A patrol car is a police officer's home for an eight hour tour (sometimes longer). Here he or she must live and work with as much convenience and efficiency as a chassis on wheels can provide. The radio that is standard equipment in every patrol unit has been supplemented by the cellular telephone.

There was a time when a cellular phone in a patrol car was considered a bit of an extravagance, but not anymore. Officers often need to telephone the station for information too routine or too sensitive to go out over police radio channels, contact citizens at home or work (saving time that would be wasted by a personal trip), call senior officers for instructions or advice, or conduct surveillance operations. All of this requires a telephone and there may not be a pay phone around (or in working order) when it's needed. It's also much safer to phone from the protection of a vehicle than it is to stand in a phone booth on a street corner..

You can identify which units are equipped with cellular mobile telephones by looking for a short black or silver "corkscrew in the center" antenna mounted on the vehicle's roof or rear trunk area. Be careful that you don't mistake an 800 MHz communications antenna for a cellular (800 MHz antennas tend to be silver with a corkscrew whip). If an officer carries a handheld cellular, an outside antenna won't be apparent.

If you don't care about legal scruples you can try to find law enforcement phone calls by searching the cellular range, which runs from 869 MHz to 894 MHz, although newer scanners don't cover these segments because monitoring them is now illegal. Somewhere among the three million and one conversations taking place you might find a cop on his mobile phone. As you listen the conversation may drop off and disappear as the vehicle moves from one area of reception into another and the call is handed off from cell site to cell site. If this happens continue scanning the band and you may pick it up again. Be aware, however, that the Electronic Communications Privacy Act of 1986, makes it illegal to listen to cellular telephone conversations of any type. If you should choose to do so anyway, even after this warning, don't call me to bail you out of jail. I provide this for informational purposes only.

UNDERCOVER OPERATIONS

One of the most fascinating aspects of law enforcement radio communications are those that take us behind the scenes of police operations. I am frequently asked how to find and monitor undercover missions. Do they use radios? Are their transmissions encrypted? What frequency ranges are used?

Undercover operations come in many different forms, such as stake-outs, moving surveillance, drug buys/busts, prostitution enforcement, and gang enforcement. Communications during undercover details are very important from an officer safety standpoint and for the ultimate success of the operation. The radio provides instant contact between officers should help be required.

The first place to look for undercover details are your local department's Tactical and Car-to-Car frequencies. These are the most accessible channels for police to use because every radio has access to them and every officer is assigned a radio. Tactical channels that are normally repeater systems may be used in simplex mode for sensitive operations.

Next, listen in on Mutual Aid or Interagency/Intercity channels. Law enforcement personnel sometimes like to move off of department-assigned frequencies for privacy and safety reasons. Don't neglect channels allocated to local government or unused fire or city frequencies. Police surveillance communications have been heard on animal control frequencies and even on city waste disposal channels! Police departments are allowed to use *any frequency* in the radio spectrum for surveillance operations as long as the transmissions generate no harmful interference and use less than two watts of power.

Surveillance can bring both fast-paced communications and periods of dead silence. Don't give up checking for possible undercover channels simply because you don't hear anything right away. You might have tuned in when a suspect is in a residence or a store and the surveillance units can do little other than sit and watch and wait. When the suspect goes on the move the radio will come alive with direction of travel information and tactical decisions.

Officers commonly use first names or nicknames when engaged in surveillance in order to keep things simple. You won't generally hear normal unit ID's.

Most surveillance by city police takes place on a single frequency which makes listening in extremely easy. Of course, low power simplex modes of transmission mean that you'll have to be close to hear things clearly.

EMERGENCY SITUATIONS

The nature of law enforcement work brings officers into tense situations every day, as they face uncooperative, combative, and sometimes violent citizens. In addition, crimes in progress, high risk traffic stops, warrant services, and use of force situations all dictate an increased level of urgency. For officer safety considerations dispatchers often shut the Dispatch frequency down, accepting only calls from the unit or units involved in the emergency situation.

Earlier we talked about the need for officers to get their messages through, which is why brevity in the form of codes and abbreviations is used. During an emergency or high risk situation it is imperative that officers have full control over the radio, with the ability to put their message on the air without interference.

When an officer needs help he will make the need very clear when he makes his call. If he needs back-up but the situation doesn't require an emergency response, he'll simply ask for cover "no code" or "not urgent." Listen carefully to what is said and the way it is said. Often the tone of voice alone will tell you what the level of urgency is. When a cop needs help every officer in the vicinity will drop the hammer to get there and assist.

"Officer in trouble," "Officer needs help," "Signal 100," "Code A," "Code 33," "10-0," "1-74," "10-77," "11-99," "Signal 1" - whatever name it goes by it means emergency, and when it's called you'll want to be there to listen. Bear in mind that when a frequency is tied up with emergency traffic for a long period of time, routine communications, which usually take place on the Dispatch frequency, may be moved to a Tactical or back-up channel for the duration. This switch might be announced or the information might have been passed along in an earlier briefing. In any case, if you are interested in keeping an ear to the normal dispatch traffic during emergency situations, take a quick scan around.

TECHNIQUES FOR LOW-PROFILE MOBILE MONITORING

Sooner or later you will be tempted to take your scanner with you on your journeys. It's inevitable; there's nothing quite like hearing the call and seeing the flashing lights as you motor along. When you do go on the road, keep these tips in mind:

1. The use of a scanner in a vehicle is illegal in some states. Check with your local authorities or consult the penal code for your state or ordinances for your city .

2. To keep from attracting undue attention, keep your scanner out of view. Between the seats or mounted low on the dashboard should be sufficient. A nice idea is to fashion a case for the scanner that can be attached to the driver's inside door panel (if you're using a handheld).

3. Keep the volume down. No matter how exciting it may be to play Joe Friday and crank up the volume, don't. Keep it just loud enough to be comfortably heard. Discretion is everything.

4. An external speaker connected to the scanner's earphone jack and then mounted overhead or on the dashboard does wonders for audibility.

5. Cellular phone holders or sturdy cup holders make great scanner mounts.

6. If you intend to leave the scanner in the car while you park and run errands it's a good idea to cover it up. A thief will love you if you leave him a scanner with all of the local police frequencies in it! Good places to hide a scanner include the glove compartment, trunk, or in a locked box. I always advise locking up the unit, but you can also cover it with a blanket, jacket, or towel. What isn't seen won't attract the attention of thieves.

MONITORING ON FOOT

Sporting events, outdoor festivals, beach parties, picnics, camping, air shows, even a trip to the local mall - these are all places where having a scanner along might be fun. Naturally, most of this type of scanning is done on foot so a handheld scanner is mandatory. Here are a few tips for monitoring on foot:

1. Most scanners come equipped with a steel or plastic belt clip or a case which a belt can be slipped through. Either way is perfect for carrying the unit along on a walk and should be used for security and to keep your hands free. Try clipping it to your side or on a waist pack.

2. Wear a jacket or long shirt to cover the scanner if you like to be inconspicuous. Use an earphone to keep the audio from disturbing others.

3. The biggest threats to handheld scanners are moisture, heat, and shock. Keep your unit away from rain and spilled drinks as many liquids contain minerals that will corrode electronic circuits. Heat will over stress and warp delicate microprocessors and LCD (liquid crystal display) windows, damage batteries and melt plastic casings so never leave the scanner in direct sunlight or in an area of high temperature for long periods of time. Shock caused by bumping, dropping, or otherwise striking the unit while carrying it in hand or on a belt can seriously misalign internal components and ruin your whole day. So make sure your scanner is secure.

4. Carry extra batteries on the off chance that those in use are low or discharged. Even if you just put a new set in, carry extras. You never know when you may need them.

5. Try taking a camera bag along to carry extra equipment. A photographer's or fisherman's vest also works great, as does a waist pack.

> *"When you are out in the same town on a daily basis, you get to know the vehicles that are at certain places at certain times of day. The business owners' cars, the cleaning crews' car, etc. When you see a new face or a new car, you check it out. Running a license check on a vehicle is the first and easiest step to take, or just writing down the license and description of the vehicle and the time and location on some note paper. In case something turns up, you have somewhere to start looking. Things that don't belong. The hard part is getting to know what belongs. Once you know that, the rest comes easy."*
>
> - Officer Keith Schroeder
> Counter Sniper Unit
> Grand Forks, ND Police Department

8

THE LAW AND POLICE MONITORING

I have taken my scanners in my vehicle, and into stores, onto military bases, to airports, train stations, and everywhere around town and have never been stopped or questioned by authorities. This is because my state has not restricted scanner ownership by the general public. Some states restrict the use of scanners in a vehicle. Laws change, however, so you should check local regulations relating to mobile scanner monitoring before you do any. Laws are made and sometimes later repealed so it's difficult to keep abreast of developments. I can't be more clear about this: know the laws! Don't take my advice, find out for yourself what the law is in your state. It is your responsibility as a citizen and a responsible scanner owner to know and obey the laws. I take pride not only in listening to law enforcement, but also in being aware of my state's laws and obeying them.

Also, be aware of a particular law which, in one form or another, applies in all states:

"Any unauthorized person who intercepts any police radio service communication, and who divulges such communication to anyone he knows to be a suspect in the commission of a crime with the intent to help the suspect escape arrest, trial, conviction, or punishment, is guilty of a misdemeanor. This crime is punishable by a fine not exceeding one thousand dollars, or imprisonment in the county jail not exceeding six months, or both." (California Penal Code 636.5)

HOW TO FIND OUT IF MOBILE MONITORING IS LEGAL IN YOUR STATE

Your State's Attorney General's office is a good place to ask about state laws relating to scanner monitoring. A letter will usually net you copies of the pertinent information. Obtain a copy of the Electronic Communications Privacy Act of 1986, also. Attorney Frank Terranella's book *Listener's Lawbook* lists all monitoring laws in America, state by state. Is has been recently updated and is invaluable if you plan on taking your scanner along with you, especially if you plan to travel

outside your own state. It's published by the Association of North American Radio Clubs, reachable on the World Wide Web at www.anarc.org. Also drop by your local library and check out their reference section for local, county, and state law listings.

RESPONDING TO A CALL: THE LINE BETWEEN ONLOOKING AND INTERFERING

Every scanner owner who monitors police services must confront a special question sooner or later - whether or not to drive to the scene of a nearby emergency call. Police, Fire and Emergency Medical Services are some of the most highly monitored agencies among the scanning set. Given the fact that these agencies are seen going about their jobs on our own city streets day after day, sirens wailing, red and blue lights flashing, it's no wonder Public Safety monitoring draws so much interest.

The easy availability of a wide variety of VHF/UHF scanners make it easy to receive Public Safety frequencies and go behind the scenes of local agencies. Though many simply listen and have no interest in going any further, some feel the experience is not complete without traveling to the scene and seeing things firsthand.

The law in many areas of the United States makes it a clear misdemeanor to sight-see at the scene of an emergency. Local and state laws, though they vary from district to district, often provide for fines and/or arrests of individuals who interfere with emergency crews. A wise monitor checks applicable laws, either through library research or by direct communication with his state's Attorney General's office.

Scanner listeners can basically be divided into two distinct operating philosophies:

The police and fire buff listens to local frequencies, just waiting for that big call, won't think twice about driving a considerable distance to reach the scene, and will often cruise around town, following emergency crews in the hopes of seeing them in action. This person believes that he or she belongs alongside public safety officers simply because he has access to the calls over a scanner. He will often cross police lines to speak with officers or firefighters, all the while carrying his radio conspicuously.

The "professional" public safety listener views the scanner as a tool through which he or she can learn the workings of emergency agencies; he or she has a keener interest in hearing communications procedures

during a call. (What units are dispatched to which incidents? From what locations? How are the radio calls made? Are Mobile Data Terminals used? What frequencies are used and in what order? What are the call signs? The unit numbers?) The professional listener's interest remains with the radio, which is what communications are all about. Though they often carry handheld scanners along with them, they know and respect law enforcement and fire officials and stay clear of any potential or known emergency areas.

Occasionally there will be some crossover, such as when a serious incident occurs within view of the professional listener's home or location, and he walks outside to watch. This behavior is only normal - most of the neighbors or other passersby are probably doing it too. As long as he remains on his own property or out of the way, all is legal.

There is, however, a line between being an interested onlooker and an interfering buff. Like so many other things in life this is a fine line. Crowds routinely come out of the woodwork to observe activity at an emergency site and in the great majority of cases this conduct is not frowned upon by police and fire officials. It's the person who shows up with his radio at all local incidents, who gets in the way, who trespasses at crime or accident sites, and who has no good reason to be there, who gives scanner hobbyists a bad name.

If you should find yourself tempted to go to the scene of an emergency here are some basic questions to ask yourself before you do:

1. Can I arrive at the call before emergency crews and assist in any way?

2. Will showing up in the area contribute to unneeded traffic flow problems or inhibit the movement of emergency vehicles and authorized personnel?

3. What are the local laws related to sightseeing at emergency scenes? Can I get fined or even arrested for being there?

4. Do I really need to be there? Won't I receive more information simply by staying home by the scanner?

If you do find yourself on scene at an emergency there are some guidelines to follow which will help ensure that you and your scanner won't become victims or unwanted trespassers:

1. If you're in a car don't park on streets that might be used for main access to the scene by emergency personnel because this might cause unwanted congestion, restrict avenues of approach and slow down the response. Keep your vehicle well clear of any emergency areas.

2. If you're on foot be aware of the movement of vehicles and remain at a distance, clear of potential danger areas which might result from explosions, downed electrical lines, hazardous materials or gunfire.

3. Never cross yellow police/fire line tape or cone barriers. These markers are there to limit public access and protect crime scenes.

4. Always heed the instructions of police officers and firefighters. If you are asked to move or leave the area, don't argue - just do it.

5. If you are a witness or otherwise have helpful or important information about a crime or accident, make contact with a police officer at the scene to relay your information. Again, do not cross barriers or enter crime scenes - wait near a patrol unit or other emergency vehicle until an officer is free to speak with you.

As scanner listeners, out and about with our radios, we often hear privileged information over the airwaves relating to cases or incidents. The law makes it illegal to communicate anything we hear to a third party. It also makes it a punishable offense to divulge such communications to anyone known to be a suspect in the commission of a crime with the intent to help the suspect escape arrests, trial, conviction or punishment.

Resist the urge to tell others around you at an emergency scene what you just heard over the scanner. Decorum demands that you keep the volume low and the communications to yourself. If you don't, you just might be asked to leave by public safety officials. Don't abuse your freedom to gain information through a scanner. It's better to stay close to the radio and monitor the action, searching for new frequencies that might be in use. Once you gain experience you won't miss much. Honest.

As scanner listeners, we have an inside edge about what transpires during life threatening public safety situations. We should hold ourselves to a higher standard, and we should know better than to get in the way.

9

HOW YOU AND YOUR SCANNER CAN HELP LAW ENFORCEMENT

You hear sirens and flip the switch on your scanner. Police are in pursuit of armed robbery suspects, but the bad guys have given the good guys the slip somewhere around Fourth and Main. The primary unit, frustrated because he lost the vehicle, gives out a description which is broadcast citywide. You listen as a perimeter is set up and police units troll the streets hoping to find the suspects. You're frustrated too, because your city isn't safe anymore. Now you have to keep the kids inside because armed robbers might be on the loose in the area. Your frustration takes you to the window where you look across at the park. Wait a second! What color was the suspect vehicle? Red? A red Camaro? You grab a pair of binoculars and sure enough, the car parked under the trees near the garbage bins is a red Camaro. As you watch, two men get out and head north up the street. Oh, oh! They fit the descriptions of the armed robbery suspects.

These are the guys! So what do you do? The call just went out a few minutes ago. No one else around here knows anything about it because they don't have scanners. The police are still searching, but they're several blocks away. Should you call the police? Or should you just mind your own business?

Moral dilemmas such as this one can happen to those of us who listen to scanners. It's a fact that we sometimes come across information that may be of use to law enforcement; information that could prevent a crime or maybe assist in solving one. Some people do nothing. Some want to get involved and share their information, but don't know how, or whether, they should. Many are afraid of retribution or of having to testify in court. Our scanners and what we hear on them put us into ethical predicaments. Do we get involved like good citizens? Or do we look the other way and forget what we heard and saw?

It is estimated that there are only 2.1 police officers for every 1,000 citizens nationwide. Not only is this an alarming statistic, but it illustrates what an overwhelmed minority law enforcement is. Chances are

better than good that when there is a crime in your neighborhood, the men and women working to keep your streets safe may take a few minutes to arrive if they're not already in your immediate area. Budget constraints and the high cost of training and equipping the modern police officer virtually guarantees that the black hats will outnumber the blue uniforms.

The only way to reduce crime is to get involved and help your police. They can't do it all and shouldn't be expected to. Scanner listeners often hear about a crime while it's still in progress and then have an opportunity to be an extra pair of eyes for police. Let's use this unique perspective on crime and law enforcement to make a difference.

Many scanner monitors I have talked with are interested in helping but they aren't sure what information is valuable to police and what isn't. Ask any 911 dispatcher and they'll tell you many of the calls they receive aren't emergencies at all and shouldn't have been called into an emergency number.

The only times you should use 911 or your local emergency telephone number is to summon police, fire, rescue, or emergency medical assistance in a life-threatening situation, or to alert the authorities to a crime in progress. An armed robbery with the robbers still in the store, a fire, a vehicle accident involving injuries, a car jacking, a fight involving a weapon, a heart attack. Remember: if an immediate response is needed, use the emergency number.

When sheriff's deputies went in pursuit of a stolen vehicle, I heard the sirens only streets away and turned on the scanner. The driver of the vehicle had bailed out and run for it. Deputies were unable to locate him. It was 2 a.m. and through an open window I heard running footsteps! The bad guy was sprinting past my house, casting looks over his shoulder at the sheriff's unit just coming to the intersection. The deputy didn't see him, and made a U-turn to head in the opposite direction. I called 911 and told the dispatcher where to send the unit. I could have called a non-emergency number but it might have taken me longer to get through and the response might have been slower. The thought that the man was running through a residential neighborhood, possibly armed, probably looking for a place to hide, made up my mind for me about which number to use. This is considered a Crime In Progress and requires an immediate response.

What if the situation had been different and I had found the abandoned car in the morning? Certainly not a case for using 911

because the bad guy would have been long gone. My unique knowledge of what had happened the night before would allow me to alert police to the location of the stolen vehicle; a vehicle which might have evidence inside that could help to find the thief. I call that Supplemental Information.

Take another example: Perhaps the PD has put out a call about a man with a gun and has given a description of the suspect. You recall that while you were driving down Second Avenue earlier you saw the very same man get into a blue Nissan with Arizona plates. This, also, is Supplemental Information, and will help law enforcement narrow their search. It's important and you should call it in. Supplemental Information should almost always be handled by using a non-emergency number, unless the information will assist in resolving a crime in progress.

There are four telephone numbers that you should have near your phone: your local law enforcement emergency number, the non-emergency or business number, the detective division number, and the community-oriented policing officer's number (if applicable), plus any others relevant to your area or your listening targets.

Your call to a police emergency or non-emergency number is answered by a call-taker, or dispatcher, who is either a civilian or a sworn officer. All are trained to ask a pre-determined set of questions in order to obtain the most critical information about the incident you are reporting. Immediately after the call-taker answers, state the type of emergency or information you are reporting. "I'd like to report a fire" or "I have information about a robbery." State right away what the call is about so that the call-taker can determine who to send. Many times emergency personnel are dispatched while you are still on the phone. Some people recommend that, in the same initial sentence, you also tell the call-taker where the incident is located. But I have found that if you're excited, speak quickly or are unclear, you will be asked to repeat the information, which will waste time. For this reason, state only the situation you are reporting and let the call-taker take it from there.

You will be asked a series of questions. Answer them clearly and to the best of your ability. Realize that while the questions may seem dumb or repetitive to you, they are designed to extract important information. Stay on the phone with the dispatcher, answer all the questions and don't hang up until the dispatcher ends the conversation. You may be asked to stay on the line to provide updates.

After you hang up the situation might change. Perhaps the fire is out, the criminal has fled the scene or the man beating his wife has left the house and is in the street. Don't hesitate to pick up the phone and call back to provide public safety with the new information. This can be vital to the response and to officer safety. Make another call, make several calls. Whatever it takes.

Sometimes we may hear things not meant for our ears, and perhaps not legal to listen to, such as cordless or cellular telephone calls. What if you hear your neighbor talking about how he robbed the 7-11 at Tenth and Grand last night, or a baby monitor transmitter reveals that a neighbor is abusing his child. What if you want to report the information but are afraid of retribution or having to appear in court? You can still call.

Caller ID displays names and telephone numbers and enhanced 911 systems have the ability to display the caller's address as well. If you don't want to be identified, go to a pay phone and make your call from there. This method might take up vital time, though, so think about simply requesting anonymity. Tell the call-taker that you would rather not be contacted by an officer regarding this call and ask that your name and location not be broadcast over the police radio. Usually, this is enough to keep you free from further involvement. You can also flag down an officer on the street and give him your information in person, requesting anonymity.

By far the best way to stay anonymous and still give your information is by using one of many "tip lines" such as Crime Stoppers. Calls to these lines go to an undisclosed location and are not recorded. Callers are usually given a code number if their information is worthwhile and are asked to call back at a later date to confirm whether or not an arrest has been made. Rewards for tips leading to arrests are common. At no time is the caller's identity revealed or even known by others. Find out if such a line exists in your area.

Fighting crime is hard work, as our men and women in law enforcement know only too well. Any bit of information, no matter how insignificant it seems, can be valuable. Only the professionals can determine what is or isn't helpful, and then only if you tell them what you know. They welcome your assistance and your information. Don't be afraid to get involved, even if you must do it anonymously. Use your scanner for more than just entertainment. If you have information about a call pick up the phone and call it in. If you don't call you'll have to live with the knowledge that you could have made a difference.

10

A BADGE

Late night is the best time to listen, when the only people out on the streets seem to be cops and criminals. It's a shattering world out there, in the shadows and under the sodium lights, where life can be quick and tough. There are guns and drugs and desperate people who have no respect for the badge.

Look deeply and you will see the despondent, the hopeless, the scared. People living on the edge, pumped with booze and drugs and crime. You will see a side of life you've never dreamed about even in your wildest, sweat-soaked nightmares. It is into this lunacy that we send our cops, and ask them to hold back the danger so that we can sleep safely.

Front pages tell only half of the story: "UNARMED MAN KILLED BY POLICE OFFICER." The reporter gives the facts: late night, liquor store, suspicious person, two shots, ambulance, emergency room, DOA. Fast and clean. We curse a trigger happy cop and pray for a young man taken from life too early. But we weren't there, and the reporter wasn't there on that street...

Did the newspaper tell you about the feel of that night? Did it tell you that it was one a.m., no moon, a silent main street in sad, tough part of town? Did it tell you how the police officer, on directed patrol, cruised slowly around the corner, window rolled down to hear the sounds of the night, radio murmuring softly.. how his headlights swept across a man peering into the front window of a closed liquor store. Did it tell you what the radio call sounded like, so routine at first: "311 John will be checking on a transient at.." and how the voice halted and then jumped ahead as the man began to run from the patrol car. ".. ah, he's running! 311 John will be in pursuit, eastbound Fourth at Main. Send me another unit."

Did that newspaper report, so full of facts about the shooting, tell you what that cop felt as he hit the accelerator, lights and siren, and went after the man? Did that article take you inside to feel the officer's

heart pumping double-time as the suspect looked back over his shoulder and you saw the shape in his hand? Did your mind whirl as you tried to see beyond the limits of your lights, as you took your foot off the accelerator, then put it back, as you wondered what he had - would he turn around with a gun and fire at you? Would you die tonight because you had come across someone planning a robbery and he had run and you had pursued because you were a cop and that was your job?

"311 John now southbound Fifth, heading into the alley."

"314 John is at Second, thirty seconds out."

Did you feel the cool of the night air as you stopped your unit in the entrance to alley and stepped out, siren echoing as it wound down to silence, engine ticking over. Did you feel the sweat break and run under your protective vest and the scrape as your gun came clear of the holster and you started that long walk. The walk down into the alley, diffused by your lights. Did you feel your heart miss a beat, as you saw your suspect at the far end, pressed against the chain link fence, poised to climb and missing his foothold. Did you hear your voice, sounding so small and ineffective in the quiet as it said: "Stop! Police officer!"

Did you see the suspect turn, his foot come back down onto the pavement, his body pivot, his hand come up, the shadows deep and black in the hollows where your lights had no effect, where your unit's overheads created a kaleidoscope of reds and blues, unnerving you. Did you see what he had in his hand in that alley at one a.m. with a siren coming, but too far off yet, and it's just you and him and does he have a gun? Is it a knife? Or a stick? Maybe a rock? Or is he giving up? Is he holding up his hands? Or is it a gun? A cop is killed in the line of duty every fifty seven hours in this country. Will you be killed here tonight and never see your family again? Your wife, your little girl, barely seven, your son who wants to grow up to be a hockey player. Is it a gun?

One second to decide. A thousand thoughts, rules, regulations, hundreds of hours of training, all compressed into one vital second.

You shoot.

Do you feel the gun go off? The roar, the smell, the awful fall, the silence. More police cars now, rolling up behind you, but you can't speak, can't acknowledge them. Your gun is hanging in your hand, hot barrel cooling in the night air.

You shot him.

The sergeant is there now -and other officers. Commotion now, on a street that moments before was so silent in the night. Do you see them

lean over the suspect, disarm him, step back and show you his weapon: a beer bottle. Oh, God! Do you hear the roaring in your ears, bitter bile in your throat, the lights like a carnival and people coming out of the woodwork to see what happened, and radios and somewhere your mike in your hand and a voice you've never heard before, low and breathless and shaking, saying:

"311 John, Code Four, roll medics to the scene."

Did the newspaper tell you how it felt to try to sleep that night? How it felt to be a cop who feared for his life and had to shoot a man in a dirty, shopworn alley in the middle of town. Of course it didn't. The newspaper told only the facts. The words were the truth, but the person who wrote them wasn't there, facing a danger only police officers know about, in a job with little respect, long hours, tough conditions… and a badge.

We give our cops badges and guns and uniforms and tell them to go out and patrol our streets, deal with our refuse, our stupidity, our desperateness. Save our children, catch our criminals, clean up our mistakes, solve our problems. Be enforcer, protector, server, referee, social worker, medic, and confessor. Then we read newspaper reports and we judge them. We judge their decisions, their actions and their methods. We curse them and cry for those they've arrested, mistreated, or shot.

And then we ask them to go out next shift and face it all over again.

> *"Scanners have been around for a long time and it's a popular hobby here in the Tampa Bay area of Florida. As a LEO [Law Enforcement Officer] I have no problem with citizens listening in. In fact, before we went to 800 MHz trunking, we used to have citizens call in and tell us which way the bad guys ran after they heard police activity in their neighborhood. When we abandoned our 460 MHz system for the 800 trunked system, we had several complaints from citizens stating that they could no longer listen in on us. I feel much more secure knowing the bad guys cannot listen in as easily, but also feel sorry for the people who enjoyed listening in."*
>
> - Officer James Heinz
> Clearwater, FL Police Department

11

IN CLOSING

The purpose of this book has been to assist you in becoming familiar with the basics of police radio communications. As you listen and gain experience with monitoring your local law enforcement agencies, you will learn more than can ever be taught in a text such as this one. It is the hours spent with the scanner that are the most rewarding. Call signs will become familiar, as will voices. You'll begin to identify units, beats, and trouble spots. You will learn things about your city about which the ordinary citizen hasn't a clue.

I encourage you to talk with officers in your city, arrange for tours of police headquarters and communications centers, attend open houses and equipment demonstrations. Check with law enforcement public relations to find out if there are patrol ride along programs for civilians or citizen academies available. Ask if you can volunteer to assist at the station. Join or start a Neighborhood Watch group and become involved in your community crime prevention plan. The police department is only half of the crime fighting team .. you are the other half. Cooperation with law enforcement is the best way to turn the streets of the 21st century into safe places.

The brotherhood that is law enforcement very seldom allows outsiders - civilians - in. Our usual contact with a cop comes during a traffic stop or an emergency, when officers are acting in their official roles. Then we see only the uniform and badge, not the human beings beneath. And there are human beings under there, believe me! Monitoring police communications allows us a very rare look inside law enforcement, a look that few civilians ever take. Listening, we hear the routine, the tension, and the hilarity of day to day police work. We hear real life, something no reality-based television series can convey. Use your scanner wisely and learn.

12

APPENDIX

THE MIRANDA WARNING

The Miranda Admonition is the result of two Supreme Court decisions (Miranda v. Arizona 385 US 436 and Escobedo v. Illinois 378 US 478) which require that a police officer advise a person under arrest of his constitutional rights. The Miranda Warning must be given to a suspect who is in custody. The warning is not required during an investigation conversation unless the officer is in doubt, at which time he may utilize it. A suspect who is not free to move about or leave the area is usually considered to be in custody.

1. You have the right to remain silent.

2. If you give up the right to remain silent, anything you do say can and will be used in court against you.

3. You have the right to speak with an attorney of your choice before questioning and to have the attorney present during questioning.

4. If you cannot afford an attorney, one will be appointed for you by the court prior to any questioning, if you so desire. The attorney will not cost you anything, his services are free.

Officer must then ask the suspect:

1. Do you understand each of these rights that I have explained to you?

2. Having in mind and understanding your rights as I have told you, are you willing to talk to me?

SEARCH WARRANT FORMAT

In the Municipal Court, _____ Judicial District, County of _____, State of _____.

The people of the state of _____, to any Sheriff, Constable, Marshal, or Policeman in the County of _____: Proof by oral statement under oath made in conformity with Penal Code Section having been made this day to me by _____ that there is probable cause for the issuance of a Search Warrant on grounds set forth in Penal Code Section _____. You are therefore commanded to make a search at any time of the day _____, good cause having been shown therefore, of the following persons or property _____, and if you find the same or any part thereof, to retain the same in your custody, subject to order of the court as provided by law.

GLOSSARY

This glossary is intended as a guide to some of the slang and terms used by law enforcement. Check your state's penal code and related laws for complete definitions of legal terminology.

ACCESSORY - Any person who, after a felony has been committed, harbors, aids, or conceals a principal.

AGGRAVATED ASSAULT - Physical attack on a person that results in serious bodily injury.

ASSAULT - Unlawful attempt to commit a violent crime on the person of another.

ALARM - Categorized as Audible or Silent, alarms come in many forms covering residential/business perimeters, interiors, or specific areas. Silent alarms usually ring at the security services' dispatch console and are relayed to police.

ARRAIGNMENT - The first appearance of a defendant before a judge after arrest.

ARREST - The act of placing a person in custody.

BATON - Also known as a night stick or PR-24, a striking device carried by police for self-defense and subject control.

BATTERY - Any willful and unlawful use of force of violence upon the person of another.

BE ON THE LOOKOUT - (BOL or BOLO) The modern version of the All Points Bulletin.

BEAT - An area or zone of the city delineated for the purposes of assigning police patrols.

BENCH WARRANT - Issued by the court for the arrest of a defendant who fails to appear at the requested time after release on bail.

BOOKING - An administrative record of an arrest made in a police station, listing offender's name, address, description, date of birth, employer, time of arrest, offense, or name of arresting officer. Booking also includes fingerprinting and photographing of the offender.

CASE NUMBER - A separate number assigned to each individual case for which a written report was prepared.

CITATION - In lieu of a physical arrest, a citation may be issued for a misdemeanor offense under the Vehicle Code or any local ordinances. The person receiving the citation must sign it as a promise to appear in court.

CITE AND RELEASE - A discretionary procedure whereby an officer writes a citation, the person signs it as a promise to appear in court at a later date and is released.
COMPLAINT - A sworn allegation made to a court or judge that an individual is guilty of a designated offense.
COMPLAINANT - A person who makes a complaint against another, or requests a police officer by phone or in person.
CONVICTION - A guilty judgment by a court.
COVER - An additional police unit(s) to provide back-up to the primary unit on the scene.
CUSTODY - Arrest.
DETAIL - Assignment.
DIRECTED PATROL - Officer-designed patrol strategy intended to target trouble spots.
DISPOSITION - Outcome of a radio call.
EVIDENCE - Testimony, writing, material objects or other things presented to prove the existence or nonexistence of a fact.
EXCLUSIONARY RULE - Constitutional right forbidding the use of illegally seized evidence in court.
FELONY - A crime punishable by death or imprisonment in the state prison.
FIELD INTERVIEW (FI) - When an officer stops a suspicious person to question them about their activities, record their name, address, etc. and take their photograph for future reference.
FIELD TRAINING OFFICER (FTO) - An experienced officer assigned to train a rookie officer during his probationary period on patrol.
FIRST RESPONDER - Usually a police officer or firefighter; any public safety officer first called to a scene.
FLEX CUFFS - Plastic strapping used in place of steel handcuffs.
HANDHELD (HT) - Portable radio.
HOT STOP - Used by officers and dispatchers to describe a vehicle stop of a suspected or known felony suspect; also known as a felony tactical stop.
INDEX OFFENSES - FBI designed crimes used in the Uniform Crime Reports.
INFRACTION - An offense not punishable by imprisonment.
KEEP THE PEACE - A common phrase used when an officer assists a citizen in going about their lawful business when the threat of violence is possible; i.e. a girlfriend wishing to retrieve her property from a former boyfriend's house might request that an officer come along to "keep the peace."

LANDLINE - Telephone.

MALICE - An intent to annoy, harm, or injure another person in any way.

MANSLAUGHTER - Unlawful killing of a human being without malice, such as voluntary upon a sudden quarrel or heat of passion, involuntary in the commission of an unlawful act, or in the commission of a lawful act which might produce death.

MISDEMEANOR - Crime punishable by imprisonment in the state prison, fine, or imprisonment in the county jail at the court's discretion.

MISSING PERSON - Anyone who is at risk and missing is considered a missing person, as it any child who has been taken, detained, concealed, enticed away, or retained by a parent in violation of custody laws, or is missing voluntarily or involuntarily, or under circumstances not conforming to his or her ordinary habits.

MURDER - The unlawful killing of another human being. First Degree Murder is murder by means of a destructive device or explosive, use of ammunition, poison, lying in wait, torture, or any other kind of willful, deliberate or pre-meditated killing, or that which is committed in the perpetration of attempt to perpetrate arson, rape, robbery, burglary, or mayhem. Second Degree Murder - All other kinds of murder.

NOTICE TO APPEAR - When a person is issued a citation for a misdemeanor in lieu of an arrest, he/she must sign a notice to appear, promising to appear in court on the requested date.

OVERHEADS - Red, blue, and white roof mounted lights on a patrol unit.

PAROLE - Release of convicts from jail to the supervision of a court officer prior to full service of the sentence.

PATROL SUPERVISOR (also Watch Commander) - The officer (commonly a sergeant), who supervises all officers in the field and is responsible for managing all major incidents.

PENDING - A term often used to preface advisory information on radio calls yet to be dispatched and awaiting a free unit to handle, i.e. "Southwest units, pending a call on North Melrose..."

PRINCIPALS - All persons connected to the commission of a crime, whether they directly commit the act constituting the offense, or aid and abet in its commission or, not being present, have advised or encouraged its commission, are principals in any crime committed.

PROBABLE CAUSE (PC) - A condition that would lead an ordinary, prudent individual to believe that a person is guilty of a crime. A police officer must have probable cause to effect an arrest.

PROTECTIVE CUSTODY - The act, by a peace officer, of taking a person or child into custody as an act of protection against immediate harm.

PUSH BARS - Vertical fender devices on patrol units used to push disabled vehicle clear of roadways.

ROBBERY - Felonious taking of personal property that is in the possession of another, from his person or presence, against his will, accomplished through fear or force.

SEARCH WARRANT - A written order, in the name of the people, signed by a magistrate, directed to a peace officer, commanding him to search for personal property and bring it before the court. Grounds for a search warrant include: when the property was stolen or embezzled; when the property was used as the means of committing a felony; when the property is in the possession of any person with the intent to use it to commit a public offense, or in the possession of anyone to whom it may have been delivered for concealment or to prevent it from being discovered; when the property consists of any item or constitutes any evidence which tends to show a felony has been committed, or a particular person has committed a felony.

SHIFT (also Tour)- Usually an eight hour tour of duty for a police officer. Most department use the three shift system, day shift, evening shift, and swing shift.

STREET GANG (CRIMINAL) - Any organization, association, or group of three or more persons, formal or informal, having as one of its primary activities the commission of a criminal act.

SUBJECT - Person under investigation.

SUBPOENA - A written order commanding a person to appear in a court of law.

SUSPECT - A person under investigation or actively wanted for a crime.

TRAFFIC STOP (also Car Stop) - The act, by a police officer, of directing a vehicle to pull over the purposes of investigation, or issuance of a citation.

VICTIM - Any person against whom a crime has been, or is intended to be, committed.

VIOLATION - Any breach of the law.
WARNING - A verbal admonition given by a police officer.
WITNESS - Any person having knowledge of facts related to a crime.

COMMONLY USED ABBREVIATIONS

AC	Aircraft crash
AID	Accident Investigation Detail
ADW	Assault with a deadly weapon
AGE	Aid in gaining entry
AKA	Also Known As
ATC	Attempt to contact
ATL	Attempt to locate
BO	Out of order
BT	Bomb threat ("Bravo Tango")
CAD	Computer Aided Dispatch
CTO	Comprehensive Time Off
CP	Complaining party
CHP	California Highway Patrol
CP	Command Post
CPD	City (or county) property damage
CRT	Information Computer
DA	District Attorney
DB	Dead body
DC	Discovering party
DDL	Driver's license
DNA	Does not apply
DMV	Department of Motor Vehicles
DOA	Dead on arrival
DUI	Driving Under the Influence (also DWI)
EOW	End of watch/shift
FI	Field Interview
FTA	Failure to appear
FTP	Failure to pay
FTO	Field Training Officer
F&U	False and unfounded
GOA	Gone on arrival
GTA	Grand theft auto
HBD	Has been drinking
HT	Handheld Transmitter
H&S	Health and Safety Code
IA	Internal Affairs

IC	Incident Command
J	Juvenile involved
NCIC	National Crime Information Center
NFD	No further description
NFI	No further information
NR	No report
OAA	Outside agency assist
OIC	Officer in charge
OR	Own Recognisance
PAB	Police Administration Building
PC	Probable Cause
PC	Penal Code
PD	Police Department
PR	Party reporting
PSMR	Please see me regarding
QOA	Quiet on Arrival (also QA)
QT	Secrecy of location required
RO	Registered owner
SO	Sheriff's Office
ST	Settled
TA	Traffic accident
TC	Traffic collision
UL	Unable to Locate (also UTL)
VIN	Vehicle identification number
W	Female involved
WA	Warned

LAW ENFORCEMENT RADIO FREQUENCY RANGES

Use these FCC-allocated frequency ranges as parameters within which to search for police related communications:

37.020 - 37.420
39.020 - 39.980
42.020 - 42.940
44.620 - 45.040
45.060 - 45.640
45.680 - 45.860
45.900 - 46.040
154.650 - 154.950
154.965 - 155.145
155.160 - 155.400
155.715 - 156.030
156.045 - 156.240
158.730 - 159.210

162.000 - 174.000 (federal)
406.100 - 420.000 (federal)
453.025 - 453.975
460.025 - 460.550
470.000 - 512.000
806.0125 - 810.9875
811.0125 - 815.9875
816.0125 - 823.9875
851.0125 - 860.9875
856.0125 - 865.9875
866.0125 - 868.9875

NATIONWIDE POLICE COMMON FREQUENCIES

155.475   National Law Enforcement Mutual Aid Radio System
122.850   Law Enforcement Aircraft
123.025   Law Enforcement Aircraft
123.050   Law Enforcement Aircraft
123.100   Law Enforcement Search and Rescue Aircraft
866.0125 Nationwide Calling F-1
855.5125 Nationwide Tactical F-2
867.0125 Nationwide Tactical F-3
867.5125 Nationwide Tactical F-4
868.0125 Nationwide Tactical F-5
868.5125 Nationwide Tactical F-6

TRACKING AND SURVEILLANCE DEVICES

40.22000	Bumper Beeper vehicle surveillance
149.3500	Body and Room Transmitters
165.9125	Body and Room Transmitters
168.0015	Body and Room Transmitters
169.2000	Body and Room Transmitters
171.4500	Body and Room Transmitters
172.0000	Body and Room Transmitters
172.2000	Body and Room Transmitters
173.3375	Body and Room Transmitters
218.0000	Pronet Felony Tracking
218.5000	Pronet Undercover and Training
219.0000	Pronet Misdemeanor Tracking
219.9000	Pronet Testing and Repair
219.9600	Pronet Tracking
908.0000	Teletrac (freq hops at 6 sec intervals from 908-910 MHz)
49-50	FM Wireless Devices Search Range
87-108	FM Undercover Transmitters Search Range
154-156	FM Undercover Transmitters Search Range
168-175	Undercover Wires Search Range
406-420	Undercover Wires Search Range

POLICE RADAR FREQUENCIES

X-Band	10.525 GHz
K-Band	24.150 GHz
Ka-Band	33.4-36.0 Ghz

GENERAL FREQUENCY BAND ALLOCATIONS

Frequency	Allocation
30.000 - 30.560	US Government
30.560 - 31.980	Business / Industry / Forestry
31.990 - 32.000	Public Safety
32.000 - 33.000	US Government
33.000 - 33.100	Public Safety
33.120 - 33.400	Business / Petroleum
33.420 - 34.000	Fire
34.000 - 35.000	US Government
35.020 - 36.000	Business / Paging
36.000 - 37.000	US Government
37.020 - 37.420	Police / Local Govt
37.460 - 37.860	Power, Water, Pipeline
37.900 - 38.000	Highway Maint / Special Emergency
38.000 - 39.000	US Government
39.020 - 40.000	Police / Local Govt
40.000 - 42.000	US Government
42.020 - 42.940	State Police
42.960 - 43.680	Business / Paging
43.700 - 44.600	Transportation - bus, truck / Cordless base
44.620 - 45.060	State Police / Forestry Conservation
45.080 - 45.860	Police / Local Govt / Highway Maint
45.900 - 46.040	Police / Emergency
46.060 - 46.500	Fire
46.520 - 46.580	Local Govt
46.610 - 46.970	Cordless Phones (base)
47.020 - 47.400	Highway Maintenance
47.440 - 47.680	Industry / Emergency
47.700 - 49.580	Industry / Cordless handset
49.670 - 49.990	Cordless Phones (handset)
50.000 - 54.000	Amateur (6-meter)
54.000 - 72.000	Broadcast TV
72.000 - 76.000	(various)
76.000 - 88.000	Broadcast TV
88.100 - 107.900	FM Broadcast
108.000 - 117.950	Aero - VOR and ILS localizer
118.000 - 136.975	Aero - communications
137.000 - 138.000	Satellite
138.000 - 144.000	US Government - land mobile

Frequency (MHz)	Use
144.000 - 148.000	Amateur (2-meter)
148.000 - 149.900	US Government - fixed/mobile, satellite
149.900 - 150.050	Satellite - radionavigation
150.050 - 150.800	US Government - fixed/mobile
150.815 - 150.965	Auto Emergency
150.995 - 151.595	Highway / Forestry / Industry
151.625 - 151.955	Business
152.030 - 152.240	Mobile phone (Base) / Paging
152.270 - 152.450	Taxi (Base)
152.510 - 152.840	Mobile phone (Base) / Paging
152.870 - 153.725	Industry
153.740 - 154.445	Fire / Govt (mobile)
154.452 - 154.482	Industry (telemetry)
154.490 - 154.625	Industry
154.650 - 156.240	Police / Govt / Emergency / Hwy
156.025 - 157.425	Maritime
157.470 - 157.515	Auto Emergency
157.530 - 157.710	Taxi (mobile) / Business
157.770 - 158.100	Mobile phone (mobile) / Paging
158.130 - 158.460	Industry
158.490 - 158.700	Mobile phone (mobile) / Paging
158.730 - 159.210	Police / Govt / Highway
159.225 - 159.465	Forestry Conservation
159.495 - 160.200	Transportation - bus, truck
160.215 - 161.610	Railroad
160.625 - 160.950	Maritime
161.640 - 161.760	Broadcast
161.500 - 162.025	Maritime
162.025 - 174.000	US Govt - land mobile
174.000 - 216.000	Broadcast TV
216.000 - 218.000	Maritime - AMTS, coast
218.000 - 219.000	IVDS - Interactive Video & Data
219.000 - 220.000	Maritime - AMTS, ship
220.0025- 220.9975	General (trunked) - base
221.0025- 221.9975	General (trunked) - mobile
222.000 - 225.000	Amateur (1.25-meter)
225.000 - 328.600	US Government - Aero
328.600 - 335.400	Aero - ILS
335.400 - 400.000	US Government - Aero
400.000 - 406.000	US Govt - Meteorological / Space

406.100 - 420.000	US Govt - land mobile
420.000 - 450.000	Amateur (70cm) / military long-range radar
422.200 - 430.000	Land mobile
450.050 - 450.925	Auxiliary Broadcasting
451.025 - 452.025	Industry
452.050 - 452.500	Taxi / Industry / Transport
452.525 - 452.600	Automobile Emergency
452.625 - 452.950	Transportation - Trucks / Railroad
452.975 - 453.000	Relay Press
453.025 - 453.975	Local Govt / Public Safety
454.025 - 454.650	Mobile Telephone
454.675 - 454.975	Mobile Telephone Air (ground)
455.050 - 455.925	Auxiliary Broadcasting
456.025 - 457.025	Industry
457.050 - 457.500	Taxi / Industry / Transport
457.525 - 457.600	Maritime - shipboard repeater /Business - low power
457.625 - 457.950	Transportation - Trucks / Railroad
457.975 - 458.000	Relay Press
458.025 - 458.975	Public Safety / Local Govt
459.025 - 459.650	Mobile Telephone
459.675 - 459.975	Mobile Telephone Air (airborne)
460.025 - 460.550	Police / Public Safety
460.575 - 460.625	Fire
460.650 - 460.875	Business - Airport use
460.900 - 461.000	Business - Central Alarms
461.025 - 462.175	Business
462.200 - 462.525	Manufacturers / Industry
462.550 - 462.725	GMRS
462.750 - 462.925	Business (paging)
462.950 - 463.175	MED (Ambulance/Hospital)
463.200 - 465.000	Business
465.025 - 465.550	Police / Public Safety
465.575 - 465.625	Fire
465.650 - 465.875	Business - Airport use
465.900 - 466.000	Business - Central Alarms
466.025 - 467.175	Business
467.200 - 467.525	Manufacturers / Industry
467.550 - 467.725	GMRS
467.750 - 467.925	Business

Frequency Range	Description
467.750 - 467.825	Maritime - shipboard (rptr at 457.xxx)
467.950 - 468.175	MED (Ambulance/Hospital)
468.200 - 469.975	Business
470.0625- 511.8875	Broadcast TV
512.000 - 806.000	Broadcast TV
806.0125- 809.7375	General - conventional
809.7625- 810.9875	General - single channels
811.0125- 815.9875	General - trunked
816.0125- 820.9875	SMR - trunked
821.0125- 823.9875	Public Safety - trunked
824.040 - 834.360	Cellular Telephone
834.390 - 835.620	Cellular Telephone (data)
835.650 - 848.970	Cellular Telephone
849.0055- 850.9735	Aircraft Telephone
851.0125- 854.7375	General - conventional
854.7625- 855.9875	General - single channels
856.0125- 860.9875	General - trunked
861.0125- 865.9875	SMR - trunked
866.0125- 868.9875	Public Safety - trunked
869.040 - 879.360	Cellular Telephone
879.390 - 880.620	Cellular Telephone (data)
880.650 - 893.970	Cellular Telephone
894.0055- 895.9735	Aircraft Telephone
896.000 - 901.000	SMR/Business/Industry - mobile
901.000 - 902.000	Personal Communications Services
902.000 - 928.000	Amateur (33cm) / US Govt radar
928.000 - 929.000	Multiple Address Systems
929.000 - 930.000	Paging
930.000 - 931.000	Personal Communications Services - base
931.000 - 932.000	PCS
932.000 - 935.000	US Govt - low-capacity fixed links (FAA)
935.000 - 940.000	SMR/Business/Industry - base
940.000 - 941.000	Personal Communications Services - base
941.000 - 944.000	US Govt - low-capacity fixed links (FAA)
944.000 - 952.000	Broadcast Auxiliary - STL
952.000 - 960.000	Multiple Address Systems
960.000 - 1215.000	Aero navigation
1215.000 -1240.000	US Govt - GPS Radionavigation Satellite
1240.000 -1300.000	Amateur (23cm)
1240.000 -1350.000	Aero - Air Route Surveillance Radar
1350.000 -1400.000	US Govt - military tactical & air-ground data

POLICE CODES

A sampling of Signal, Incident and 10 Codes used by different police departments.

Dallas, Texas Police Dept.
Signal Codes:
DH	Drug House
2	Witness
3	Hang up call
4911	Hang up
6	Disturbance
6G	Random Gunfire
6X	Major Disturb.
7	Minor Accident
7X	Major Accident
8	Drunk
9	Theft
11	Burglary
12	Burglar Alarm
13	Prowler
14	Cutting
15	Assist Officer
16	Injured Person
18	Fire Alarm
19	Shooting
20	Robbery
21	Holdup Alarm
22	Animal Complaint
23	Parking Violation
24	Abandoned Property
25	Criminal Assault
26	Missing Person
27	Dead Person
28	Sick Person
29	Open Building
30	Prisoner
31	Criminal Mischief
32	Suspicious Person
33	Poisoning
34	Suicide
35	Emergency Blood
36	Abandoned Child
37	Street Blockage
38	Meet Complainant
39	Speeding
40	Other
41-20	Robbery in Progress
41-25	Criminal Assault
41-40	Kidnapping in Progress
41	Felony
42	Pursuit
44	Person in Danger
50	Eat
51	Coffee
52	City Court
53	County Court
54	Escort
55	Traffic Violation
56	Out at Station
57	Out at Garage
58	Routine Investigation
59	Follow-Up
60	Special Assignment
61	Foot Patrol
62	Public Service
63	Cover Element
64	Radio Shop
65	Telephone
66	End Duty Tour
67	Monitor Radio
Code 1	Normal Response
Code 3	Emergency
Code 4	Disregard
Code 5	En Route
Code 6	Arrived
Code 10	Known Offender
Code 10C	Known Dangerous
Code 10W	Felony Warrant
Code 10X	Stolen Vehicle

South Florida Police Codes
Signal Codes:

0	Armed - Caution	40	Vandalism
1	Drunk Driver	41	Robbery
2	Drunk Pedestrian	42	Child Molest
3	Hit and Run	43	Lewd - Lascivious Acts
4	Accident	44	Boat - Marine Accident
5	Murder	45	Airplane Crash
6	Escaped Prisoner	46	Bomb Threat
7	Dead Person	47	Vice Case (Non Drug)
8	Missing Person	48	Open Door
9	Stolen Tag	49	Alarm
10	Stolen Car	50	Labor Trouble
11	Abandoned Vehicle	51	Trespassing
12	Reckless Driver	52	Forgery
13	Suspicious Vehicle	53	Embezzlement
14	Information	54	Hijack
15	Special Detail	55	Explosion
16	Child Abuse	56	Beverage Violation
17	Contact	57	Narcotics
18	Felony	58	Unlawful Assemblies
19	Misdemeanor	59	Fire Bombing
20	Mentally Ill Person	60	Sniper Fire
21	B & E	61	Gambling
22	Disturbance	62	Racial Trouble
23	Pedestrian - Hitchhiker	63	Impersonating Officer
24	Kidnapping	64	Larceny - Auto Parts
25	Fire	65	Shoplifter
26	Drowning	66	Civil Matter
27	Prowler - Peeping Tom	67	Sick or Injured Person
28	Civil Disorder	68	Police Service Call
29	Reckless Operation	69	Loose Farm Animal
30	Larceny	70	Animal Bite
31	Assault	71	Snake
32	Suicide	72	Lost or Found Property
33	Shooting		
34	Stabbing	10 Codes:	
35	Rape	10-1	Receiving Poorly
36	Disturbance - Fight	10-2	Receiving Well
37	Disturbance - Juvenile	10-3	Stop Transmitting
38	Disturbance - Domestic	10-4	OK
39	Disturbance - Neighbor	10-5	Relay to…

10-6	Busy	10-49	Serving Warrant
10-7	Out of Service	10-50	Stopping Vehicle
10-8	In Service	10-51	En route
10-9	Repeat	10-52	ETA
10-10	Out of Service / On Call	10-53	Coming by Office
10-11	Say Again	10-54	Negative
10-12	Visitors Present	10-55	Car to Car
10-13	Weather Road Report	10-56	Meet
10-14	Convoy or Escort	10-57	At Pistol Range
10-15	Prisoner in Custody	10-59	Teleprinter Message
10-16	Pick up Prisoner	10-60	Assist Motorist
10-17	Conduct Investigation	10-61	Request Sergeant
10-18	Complete Quickly	10-62	Request Lieutenant
10-19	Return to Station	10-63	Request Bomb Squad
10-20	Location	10-64	Request Crime Lab
10-21	Call Station by Phone	10-65	Can You Copy
10-22	Disregard	10-66	Cancel
10-23	Standby	10-67	Serving Civil Papers
10-24	Trouble - Send Help	10-68	Request Legal Advice
10-25	In Contact with	10-69	Request Narcotics Off.
10-26	Message Received	10-70	Request Wrecker
10-27	Complete Record Chk	10-71	Request Ambulance
10-28	Check Registration	10-72	Request Aviation Unit
10-29	Stolen or Wanted	10-73	Request Juvenile Unit
10-30	Against Rules	10-74	Recall
10-31	In Pursuit	10-89	On Page
10-32	Is DO Operator Avail?	10-90	Scramble Code
10-33	Emergency Traffic	10-91	Tactical Frequency
10-34	Jail Break	10-92	Information Frequency
10-35	Confidential Info	10-93	Request Detective
10-36	Time	10-94	Request Back-up
10-37	Operator on Duty	10-95	Computer Check
10-38	Block Roads at	10-96	Radio Telephone Patch
10-39	Message Delivered	10-97	Arrived at Scene
10-40	Meal	10-98	Completed Assignment
10-41	In Possession of	10-99	NCIC/FCIC Hit
10-42	Out of Service at Home	10-100	Alert Keep in Contact
10-43	Any Traffic for	10-101	Log on Digital
10-44	Pick up Papers at	10-102	Log off Digital
10-45	Call by Phone	10-103	Computer down
10-46	Urgent	10-105	Prisoner Transport
10-47	Blood Run	10-106	MedicalAssist Request
10-48	End - Received	10-107	CRU Requested

Mesa, Arizona, Police Dept.

Disposition Codes:

DC1	Daily Report to be Written	10-40	Stolen Vehicle
DC2	Gone on Arrival (GOA)	10-42	Officer at Home
DC3	Unfounded	10-45	Meet with at
DC4	Message Delivered	10-52	Switch to Channel #
DC5	Civil Matter, No Further	10-53	Locate
DC6	False Alarm	10-71	At Police Range
DC7	Traffic, Crowd Control	10-78	Juvenile Detention
DC8	Field Investigation	10-79	Death Message
DC9	Turned over to other	10-90	Traffic Signal Out
DC10	Gas Drive Off	10-91	Motorist in Trouble
		10-97	Arrived at Scene
		10-98	Finished with Last

10 Codes:

10-01	Receiving Poorly		
10-02	Receiving Well	Dispatch Codes:	
10-03	Emergency Traffic	000	No Code
10-04	Acknowledgment	101	Woman
10-05	Stake Out	105	Out of Service, fuel
10-06	Busy or Out with ...	106	Car Wash
10-07	Out of Service	211	Armed Robbery
10-08	In Service	234	Check for Marijuana
10-09	Repeat Traffic	235	Check for Drugs
10-10	Out of Service	237	Check for Glue Sniffing
10-11	Out for Restroom	238	Trespassing
10-14	Transporting Subject	239	Fight in Progress
10-15	Prisoner in Custody	240	Assault
10-19	Return to Station	245	Assault w/Deadly
10-20	Location	261	Rape
10-21	Telephone	265	Prostitution
10-22	Cancel, Disregard	300	Gambling
10-23	Stand By	310	Molest
10-25	Bar Check	311	Indecent Exposure
10-27	Driver's License Check	315	Forgery
10-28	Registration Check	316	Bogus Check
10-29	Wants and Warrants	318	Defrauding Innkeeper
10-30	Does not Conform	319	Lost Property Report
10-33	Emergency Traffic	320	Found Property Report
10-35	Confidential	390	Drunk
10-36	Time Check	412	Obscene Phone Call
10-37	Prepare to Copy	413	Harassment
10-38	Warrant Confirmation	414	Threat

Code	Description	Code	Description
415	Neighbor Trouble	804	Dog Disturbing
416	Criminal Damage	805	Miscellaneous Animal Incident
417	Juveniles Disturbing		
418	Drunks Disturbing	872	Dispatch Call Sign
419	Noise	900	Welfare Check
420	Domestic Violence	901	Sick or Injured Person
421	Shots Fired	901S	Shooting
422	Gun (Subject with Gun)	902	Stabbing
423	Knife	903	Dead Body
424	Civil Matter	904	Ambulance Dispatched
437	Minor in Possession	905	Fire Follow Up
441	Drive By Shooting	906	Officer Assistance
451	Homicide	907	Follow Up Investigation
459	Burglary	908	Check Solicitor
459A	Burglary Alarm	909	Other Agency Assist
487	Theft	916	CityCode Violation
488	Shop Lifting	917	Abandoned Vehicle
491	Kidnapping	918	Insane Person
507	Traffic Control	921	Prowler
508	Traffic Congestion	924	Rescue Truck
509	Stolen Vehicle	925	Ambulance Request
509R	Stolen Vehicle Rec.	926	Wrecker Request
510	Speeding Cars	927	Unknown Trouble
585	Traffic Hazard	955	Fire Truck Request
586	Illegal Parking	956	Identification Division Requested
600	Information Needed		
601	Missing Person	961	Accident (No Injuries)
602	Child Neglect	962	Accident (Injuries)
646	Suspicious Activity	963	Accident (Fatality)
647	Suspicious Person	964	See the Party
648	Suspicious Vehicle	990	Wagon Required
666	Suicide	998	Officer Involved Shoot
692	DUI	999	Officer Required
693	Reckless Driving		
707	Bomb Threat		
711	Intensive Patrol		
801	Animal Pickup		
801D	Dead Dog in Road		
802	Cat Pickup		
802D	Dead Cat in Road		
803	Animal at Large		

San Francisco, CA Police Dept.
10 Codes:

10-1	Receiving Poorly	240	Assault/Battery
10-2	Receiving OK	245	Aggravated assault
10-4	Message Received	261	Sexual assault/rape
10-7	Out of service at	288	Sexual assault, under 15
10-8	FIn service (Foot patrol)	311	Indecent exposure
10-9	Repeat	400	Demonstration/Pickets
10-13	Advise conditions	404	Riot
10-14	Escort	405	Citizen holding prisoner
10-20	Location	406	Officer needs help
10-22	Cancel	407	Prisoner transportation
10-23	Standby	408	Ambulance
10-25	Respond as back-up	409	Tow truck
10-28	Registration	410	Assistance Responding
10-29	Check for wanted	415	Noise
10-30	Wanted person/vehicle	416	Citizen standby
10-31	Person has arrest record	417	Person ringing doorbells
10-32	Not wanted	418	Fight or dispute
10-34	Confidential	419	Fight with weapons
10-35	Consent to search	420	Juvenile Disturbance
10-36	Emergency notification	459	Burglary
10-43	Intelligence division	470	Fraud
10-96	Undercover	487	Grand theft
10-97	Arrived at scene	488	Petty theft
10-98	En route to new location, present assignment	496	Stolen property, receiving
		518	Non-injury accident
		519	Injury accident
Incident Codes:		520	Aided case
100	Alarm (Audible or Silent)	527	Bonfire
148	Resisting Arrest	528	Fire
152	Drunk Driver	529	Explosion
187	Murder/Homicide	530	Bomb threat
207	Kidnapping	531	Suspected explosive device found
211	Robbery		
212	Strong arm Robbery	585	Traffic stop
213	Purse snatch	586	Traffic congestion
216	Shots fired	587	Parking violation
217	Shooting/Attempted	588	Driveway violation-Tow
219	Stabbing/Cutting	594	Malicious mischief
221	Person with a gun	595	Graffiti
222	Person with a knife	596	Abandoned vehicle

600	Roadblock	\multicolumn{2}{l	}{Los Angeles, CA Police Dept.}

600	Roadblock	Code 1	Acknowledge Call
601	Tresspasser	Code 2	Routine Call
602	Person breaking in	Code 2	High Priority Call
603	Prowler	Code 3	Emergency Call
647B	Prostitue	Code 4	No Further Assistance
650	Threats	Code 5	Stakeout
799	Senile person	Code 6	Out of Car
800	Mentlly disturbed person	Code 6A	Out of Car / Assist
801	Person attempting suicide	Code 6C	Out of Car Suspect Wanted
802	Coroner's case		
806	Juvenile beyond control	Code 6G	Out of Car on Gangs
807	Missing juvenile	Code 7	Meal Break
809	Missing person	Code 8	Fire Alarm
811	Intoxicated person	Code 8A	Working Fire
819	Rolling intoxicated person	Code 10	Request to Clear Frequency
851	Stolen vehicle		
852	Auto boost/strip	Code 12	False Alarm
853	Recovered stolen vehicle	Code 20	Notify Media
901	Call station	Code 30	Burglar Alarm
902	Return to station	Code 37	Stolen Vehicle
904	Meet with officer	Code 77	Possible Ambush
905	Meet w/city employee	Code 99	Emergency
909	Citizen requests inteview	Code 100	In Position
910	Check on well-being	148	Resisting Arrest
911	Broken window	187	Homicide
912	Person dumping rubbish	207	Kidnapping
913	Complaint unknown	211	Armed Robbery
914	Person down	240	Assault
916	Suspicious person-vehicle	242	Battery
917	Suspicious person	245	ADW
918	Person screaming for help	246	Shooting at Dwelling
5150	Insane person	261	Rape
		288	Lewd Conduct
		311	Indecent Exposure
		374	Illegal Dumping
		390	Intoxicated Person
		415	Disturbance
		417	Gun
		451	Arson
		459	Burglary
		470	Forgery

480	Hit and Run	10-16	Domestic problem
484	Theft	10-17	Armed robbery
487	Grand Theft	10-18	Quickly
488	Petty Theft	10-19	Return to_____
502	DUI	10-20	Location
		10-21	Call_____ by telephone

10 Codes:

10-10	Out of Service	10-22	Disregard
10-11	Talk Slower	10-23	Arrived at scene
10-12	Visitors Present	10-24	Completed assignment
10-14	Escort	10-25	Report in person
10-15	Enroute With Prisoner	10-26	Detaining subject
10-16	Pick Up Prisoner	10-27	Driver's license info
10-17	Pick Up Document	10-28	Vehicle registration
10-23	Stand-By	10-29	Check stolen/wanted
10-24	Trouble at Station	10-30	Unnecessary radio use
10-27	Check Driver's License	10-31	Crime in progress
10-34	Resume Broadcast	10-32	Subject with firearms
10-35	Confidential Info	10-33	Emergency
10-36	Correct Time	10-34	Riot
10-37	Name of Operator	10-35	Transporting liquor
10-39	Message Delivered	10-36	Correct time
10-86	Traffic Check	10-37	Suspicious person
10-97	Arriving On Scene	10-38	Stopping suspect .
10-98	Assignment Complete	10-39	Resume operations
10-99	Emergency	10-40	Not stolen
		10-41	Beginning tour of duty
		10-42	Ending tour of duty

State of Georgia Police Codes

10 Codes:

10-0	Use caution	10-43	Murder reported
10-1	Unable to copy	10-44	Suicide or attempt
10-5	Relay	10-45	Hold evidence
10-6	Busy, unless urgent	10-46	Assist motorist
10-7	Out of service	10-47	Emergency road repair
10-8	In service	10-48	Traffic light out at _____
10-9	Repeat	10-49	Speeding auto
10-10	Fight or disorder	10-50	Accident
10-11	Dog case	F	Fatal
10-12	Stand by	PI	Personal Injury
10-13	Weather	PD	Property Damage
10-14	Prowler report	10-51	Wrecker needed
10-15	Burglary	10-52	Ambulance needed
		10-53	Road blocked at _____
		10-54	Livestock on road

10-55	Intoxicated driver		10-96	Mental subject
10-56	Intoxicated pedestrian		10-97	Check signal
10-57	Hit and run		10-98	Jail break
10-58	Direct traffic		10-99	Wanted/stolen indicated
10-59	Convoy or escort			
10-60	_____will leave this station at_____hours		**New York City Police Dept.**	
			10-1	Call Your Command
10-61	FBI signal in following cars		10-2	Report To Command
10-62	Reply to message		10-3	Phone Dispatcher
10-63	Prepare to make copy		10-4	Acknowledgment
10-64	Warrant		10-5	Repeat Message
10-65	Mechanical breakdown		10-6	Stand By
10-66	Checked schools		10-7	Verify Address
10-67	Trouble with individual		10-10	Possible Crime
10-68	Dispatch information		10-11	Alarm
10-69	Message received		10-12	Holding Suspect
10-70	Fire		10-13	Assist Police Officer
10-71	Homicide reported		10-14	Occupied and Suspect
10-72	Coroner needed		10-15	Verify if Stolen
10-73	Investigator needed		10-16	Vehicle is Stolen
10-74	Negative		10-17	Vehicle is Not Stolen
10-75	In contact with_____		10-20	Robbery
10-76	En route to_____		10-21	Burglary
10-77	ETA		10-22	Larceny
10-78	Need assistance		10-24	Assault
10-79	Notify coroner		10-29	Other Crime in Past
10-80	Chase in progress		10-30	Robbery In Progress
10-81	Give location and status		10-31	Burglary In Progress
10-82	Reserve lodging		10-32	Larceny In Progress
10-83	Work school crossing at		10-33	Report of Explosive
10-84	Special detail		10-34	Assault In Progress
10-85	Delayed due to_____		10-35	Other Crime In Progress
10-86	Officer/operator on duty		10-45	Rapid Mobilization
10-87	Pickup/distribute checks		10-46	Rapid Mobilization
10-88	Present telephone number		10-47	Same as above
10-89	Bomb threat		10-48	Other Rapid Mobilize
10-90	Bank alarm at_____		10-50	Disorderly
10-91	Pickup prisoner/subject		10-51	Roving Band
10-92	Improperly parked vehicle		10-52	Dispute
10-93	Blockade		10-53	Vehicle Accident
10-94	Drag racing		10-54	Ambulance Case
10-95	Prisoner in custody		10-55	Ambulance Case

10-56	Is Ambulance Needed?
10-57	Ambulance (2nd Call)
10-58	Assist A
10-59	Fire Alarm
10-61	Precinct job
10-62	Out Of Service
10-63	Out on Meal
10-65	Utility Truck Response
10-66	Unusual Collapse
10-67	Traffic
10-80	Cancel
10-82	Out of Service w/arrest
10-83	On Paperwork, Station
10-84	Arrived At Scene
10-85	Need Additional Unit(s)
10-86	Female
10-87	Unit to Hospital
10-89	Other Interim Status
10-90N	Notice Served on unnecessary job
10-90U	Unable to Gain Entry
10-90X	Unfounded
10-90Y	Unnecessary
10-90Z	Gone On Arrival
10-91	Non-Crime Corrected
10-92C	Felony/Misdemeanor
10-92Q	All Other Arrests
10-93	Report Prepared ('93C' when Crime Reported)
10-94	Handled by Previous
10-95	Non-Crime Referred
10-96	Summons Served
10-97	Aided Case Completed
10-98	Resuming Patrol
10-99	Other Final Disposition

Ohio State Highway Patrol

Signal Codes:

1	Out of service
2	In service
3	Out of service, subject to call
4	Out of service, equipment
5	Rush
6	Reference previous traffic
7	At your convenience
8	Unable to read
9	Unable to answer at this time
10	Call given point by phone
11	Call GHQ by phone
12	Call DHQ by phone
13	Call your post by phone
14	Call radio by phone
15	Call home by phone
16	Technical problems with computer
17	Accident report number
18	Have traffic - Relay
19	Relay by phone
20	Contact given place in person
21	Contact GHQ in person
22	Contact DHQ in person
23	Contact your post in person
24	Contact radio in person
25	Contact home in person
27	Improper procedure
29	Relay in person
30	Accident reported fatal
31	Accident, property damage
31A	Accident, personal injury
32	Traffic jam or road blocked
33	Drowning
34	Fire motor vehicle
36	Fire other than motor vehicle
37	Disabled motor vehicle
38	Eating
39	Location
40	Emergency
41	Immediately

42	Be on station
43	All units, call your post
44	All units, report to your post
45	Resume regular patrol
46	FCC inspector on station
47	Traffic preference
48	Stand-in examinee
49	Forced landing
50	Unidentified plane
51	Plane crash
52	Radioactive materials
53	Hazardous load
54	Wrecked truck, explosives
55	Explosion
56	Train accident
57	Bomb scare
58	Suspect can hear radio traffic
59	Permits
60	Convoy or reference convoy
75	Wanted, felony
76	Wanted, misdemeanor
77	Wanted and warrants check
79	Mentally disturbed person
88	Officer needs help

Birmingham, Alabama Police Dept.

10-00	Emergency
10-01	Ambulance Clearance
10-02	Ambulance Needed
10-03	Paramedic
10-04	Acknowledge
10-07	Eating
10-08	In Service
10-09	Repeat
10-10	Out of Service
10-11	Negative
10-14	Correct Time
10-15	Have Prisoner in Custody
10-16	Wagon Needed
10-18	Any Messages?
10-20	Current Location
10-21	Call by Telephone
10-21a	Call for Number
10-21h	Call Home
10-21s	Call Spouse
10-22	Report in Person to
10-23	Arrived at Scene
10-24	Situation under Control
10-25	See a Party
10-28	Vehicle Registration Info
10-29	Check for Wanted
10-33	Emergency: Priority
10-41	Beginning Tour of Duty
10-42	Ending Tour of Duty
10-44	Leaving Assigned Patrol
10-45	Out of Service - No Calls
10-46	Assist Motorist
10-48	Meet an Officer
10-49	Traffic Device Repair
10-51	Wrecker Needed
10-59	Blood Run
10-61	Return to
10-68	Information Needed
10-72	Supervisor Needed
10-74	Intoximeter Operator
10-75	Evidence Tech Needed
10-87	ETA
10-100	Hot Pursuit

Disposition Codes:
A - Incident Report, No Arrest
B - Arrest Report, No Incident
C - Accident Report Made
D - Follow up - Supplement
E - Incident Report and Arrest
F - Accident Report and Arrest
G - Accident Report and Incident
H - Accident /Arrest/Incident Report Made
I - False Alarms - Report Made
J - Traffic Citation
K - Assisted a Citizen
L - Assisted Another Unit
M - Cancelled by Dispatcher
N - No Report (Unable to Locate Incident or No Report Needed)

Signal Codes:
- 00 Officer Needs Help
- 01 Burglar Alarm
- 02 Burglary/theft in Progress
- 03 Rape in Progress
- 04 Prowler
- 05 Robbery Alarm
- 06 Robbery in Progress
- 07 Non-verbal E911 Call
- 08 Theft Of/From Vehicle in Progress
- 09 Homicide/death
- 11 Sexual Assault/rape
- 12 Robbery
- 13 Assault
- 14 Abortion
- 15 Kidnapping
- 20 Arson/fire
- 20b Bomb Threat
- 20f Explosive Found
- 21 Extortion
- 22 Burglary
- 23 Larceny/theft/ubev
- 24 Stolen Vehicle
- 24r Recovered Stolen Vehicle
- 50 Obstructing Justice
- 51 Bribery
- 52 Weapons Offense
- 53 Disorderly Person
- 54 Traffic Offense
- 56 Parking Violation
- 57 Trespassing
- 58 Smuggling
- 25 Forgery
- 26 Fraud/theft of Services
- 27 Embezzlement
- 28 Stolen Property
- 29 Damaged Property
- 35 Narcotics Complaint
- 36 Sex Offense/Indecent
- 37 Obscene Material
- 39 Gambling
- 40 Commercialized Sex
- 41 Liquor
- 42 Public Drunkenness
- 48 Obstructing Police
- 49 Fight/Escape/Wanted Person
- 61 Tax/Revenue
- 62 Conservation
- 63 Vagrancy/Loitering
- 64 Suspicious Person/Vehicle
- 65 Hazardous Road Condition
- 66 Traffic Accident
- 67 Direct Traffic/Escort
- 68 Assist Citizen
- 70 Crime Against Person
- 71 Unclassified Property Crime
- 72 Morals/Decency Crime
- 73 Public Order Crime
- 74 Loud Party
- 77 Person Down
- 78 Mentally Disturbed Person
- 79 Runaway/Missing Person
- 80 Airport Emergency
- 90 Desk Assignment
- 99 Miscellaneous Complaint

State of Florida 10 Codes
- 10-14 Unit Check
- 10-15 Prisoner in Custody
- 10-16 Transporting Prisoner
- 10-17 Continuing Invest.
- 10-18 Urgent
- 10-19 Return to Station
- 10-20 Location
- 10-21 Phone Station
- 10-22 Disregard
- 10-23 Standby
- 10-24 Emergency Backup
- 10-25 Contact
- 10-26 Message Received
- 10-27 Drivers License Check
- 10-28 Registration Check
- 10-29 Wants or Warrants

10-29P	Wanted Person Check	10-97	Arrived At Scene
10-30	Against Rules	10-98	Assignment Complete
10-31	In Pursuit	10-99	NCIC/FCIC/Local Hit
10-32	Breathalyzer Operator	10-99A	Hit-Article
10-33	Emergency Traffic	10-99F	Felony Hit-Person
10-35	Confidential	10-99M	Misdemeanor Hit-Person
10-36	Time	10-100	Alert
10-37	Operator On Duty		
10-38	Routine Backup	Dispatch Signals:	
10-39	Message Delivered	0	Armed Subject
10-41	In Possession of _____	1	Reckless Driver
10-42	At Home	2	Intoxicated Subject
10-43	Information	3	Hit and Run Crash
10-44	Paperwork	3	PI Hit and Run/Injuries
10-45	Phone	4	Motor Vehicle Crash
10-48	End of Message	4	PI " " Injuries
10-49	Serving Papers	5	Homicide
10-50	Traffic Stop	6	Escaped Prisoner
10-51	En route	7	Death Investigation
10-52	ETA	8	Missing Person
10-54	Negative	8C	Missing Child
10-55	Assign Channel	9	Lost/Stolen Tag
10-56	Meet	10	Stolen Vehicle
10-57	Teletype Channel	11	Abandoned Vehicle
10-63	Meal Break	12	Traffic Problem
10-64	Personal Break	12P	Parking Problem
10-65	Copy Call/Information	13	Suspicious Activity
10-66	Cancel	13H	Incomplete 911 Call
10-68	Request Evidence Tech.	13P	Suspicious Person
10-69	Request Crime Scene	13U	Unsecured Condition
10-85	Shift Change	13V	Suspicious Vehicle
10-86	Starting Duty	14	Information
10-87	Ending Duty	15	Special Detail
10-88	Advise Phone Number	16	Miscellaneous
10-89	Available On Pager	17	Follow-up Investigation
10-90	At Police Academy	18	Beverage Violation
10-91	At ACDC	19	Narcotics Violation
10-92	At Court/Deposition	20	Mentally Impaired
10-93	In Shift Briefing	21	Burglary
10-94	In Training	21B	Burglary-Business
10-95	Vehicle Maintenance	21C	Burglary-Conveyance
10-96	At Station	21R	Burglary-Residence

22	Disturbance	43L	Lost Property
22C	Disturbance-Civil	44	Juvenile Problem
22F	Physical Fight	45	Shots Heard
22T	Trespass	46	Trash Dumping
22V	Disturbance-Verbal	47	Harassing Phone Calls
23	Riot/Civil Disturbance	47	Death Threats
24	Robbery	48	Wanted Person
24A	Robbery-Armed	49	Serving Warrant
24S	Robbery-Strong Armed	50	Assistance Call
25	Fire	50A	Assist Other Agency
25A	Arson	50C	Assist Citizen
26	Drowning	50D	Assist Disabled Vehicle
27	Prowler	51	Noise Complaint
28	Sex Offense	52	Increased Patrol
29	Assault/Battery	53	Animal Problem
29X	Sexual Battery	55	Vice Violation
30	Officer Needs Help	56	Forgery/Counterfeiting
31	Medical Emergency	60	Hazmat Incident
32	Theft	61	Hazardous Conditions
32S	Shoplifter In Custody		
33	Alarm	**New Mexico State** Codes	
33A	Audible	10-10	Out of service (food)
33B	Burglar	10-11	Dispatching too fast
33F	Fire	10-12	Visitors present
33L	Lifeline	10-13	Advise weather
33P	Panic	10-14	Convoy or escort
33R	Robbery	10-15	Prisoner in custody
34	Bomb Threat	10-16	Pick up prisoner at_____
34D	Explosive Device	10-17	Traffic hazard on highway
35	Person Shot	10-18	Livestock on highway
36	Person Stabbed	10-19	Return to your station
37	Suicide	10-20	Location
37A	Suicide Threat/Attempt	10-21	Call this station by phone
38	Hostage/Abduction	10-22	Take no further action
39	Marine Incident	10-23	Stand by
40	Aircraft Incident	10-24	All units report to ___
41	Recovered Property	10-25	Do you have contact
42	Criminal Mischief	10-26	Do not use siren/lights
43	Property Call	10-27	Any answer reference
43D	Damaged Property	10-28	Check registration/DL #
43F	Found Property	10-29	Check for wanted

10-30	Does not conform to regulations.
10-31	Bomb scare
10-32	Demonstration
10-33	Emergency traffic
10-34	Clear for local dispatch
10-35	Confidential information
10-36	Correct time
10-37	Operator
10-38	Send mechanic to_____
10-39	Civil Defense dispatch
10-40	Progress on assignment
10-41	Female in patrol unit
10-42	Officer_____at his house
10-43	Drag racing at_____.
10-44	Accident, no inj.
10-45	Accident with injuries
10-46	Wrecker requested at_____
10-47	Drunk driver
10-48	Use caution
10-49	Any traffic for this unit?
10-51	Industrial accident
10-52	Drowning at ___
10-50	No traffic
10-53	Officer is clear
10-54	Have car stopped, may be dangerous
10-55	Ambulance requested
10-56	Change location
10-57	Drunk pedestrian
10-58	Mental patient (violent)
10-59	Mental patient (nonviolent)
10-60	Emergency assistance needed at_____
10-61	This officer has been injured
10-62	Police unit in accident
10-63	Dispatch coroner to ___
10-64	Domestic problem
10-65	Clear for message
10-66	Clear for cancellation
10-67	Station ____carry message
10-68	Fight
10-69	Breaking & entering
10-70	Crime in progress
10-71	Homicide
10-72	Place roadblock at_____
10-73	Lift roadblock at_____
10-74	Rape
10-75	Stolen vehicle
10-76	En route
10-77	Mobile unit switch
10-78	Prowler
10-79	Deceased person
10-80	Armed and dangerous
10-81	Officer will be at station
10-82	Make reservations
10-83	Undercover investigation
10-84	Informant in unit
10-85	Surveillance only
10-86	Attempted suicide
10-87	Meet officer_____at _____
10-88	Advise phone # for call
10-89	Assault
10-90	Police aircraft going down
10-91	Police aircraft emergency
10-92	Police aircraft closing flight plan
10-93	Police aircraft advise WX
10-94	Switching to FAA frequency
10-95	On ground and secured
10-96	Campus unrest
10-97	Arrived at scene
10-98	Last assignment complete
10-99	Officer needs assistance
10-100	Riot conditions exist
10-101	Investigate disturbance

Incident Codes:

Code 1	Murder or manslaughter
Code 2	Rape
Code 3	Robbery
Code 4	Assault
Code 5	Burglary
Code 6	Theft
Code 7	Auto theft
Code 9	Forgery
Code 10	Fraud/embezzlement
Code 11	Weapons, firing, etc.
Code 13	Prostitution/vice
Code 14	Indecent exposure, etc.
Code 15	Offenses against family
Code 16	Narcotics violations
Code 17	Violations liquor laws
Code 18	Drunk
Code 19	Disorderly conduct
Code 20	Vagrants
Code 21	Gambling
Code 22	DWI
Code 23	Reckless driving
Code 24	Suspicious person
Code 26	Arson/vandalism
Code 27	Open business
Code 28	Frequent patrol
Code 29	Dog bites
Code 30	Suicide
Code 31	Miscellaneous deaths
Code 34	Fight in progress
Code 35	Bank alarm

Ottawa Carleton, Canada Regional Police

10-22	See Complainant
10-23	Indecent Act/Exposure
10-24	Person Using Firearm
10-25	Detaining Subject
10-27	Drowning
10-28	Vehicle Registration
10-29	Check for Wants
10-30	Improper Use of Radio
10-31	Sickness/Injured Person
10-32	Obstruction/Tow
10-33	911 Activation
10-34	Alarm-Other
10-35	Major Crime Aler
10-36	Correct Time
10-37	Theft
10-38	Shoplifting
10-39	Purse Snatch
10-40	Stolen/Recovered
10-41	Break & Enter
10-42	Armed Robbery
10-43	Sexual Assault
10-44	Homicide
10-45	Sudden Death
10-46	Suicide
10-47	Fraud
10-48	Convoy or Escort
10-49	Threats/Obscene Calls
10-50	MVA
10-51	Tow Truck
10-52	Ambulance
10-53	Supervisor
10-54	Identification Unit
10-55	Domestic Disturbance
10-56	Missing Person
10-69	Are You Alone?
10-70	Subject Probationary
10-71	Subject Refused
10-72	Subject Prohibited
10-73	Subject Elopee
10-74	Subject Is Missing
10-75	Subject in Pointer Person Category
10-76	Narcotic Control Act
10-77	Emergency Landing
10-79	Hostage - Extortion
10-80	Special Situation
10-81	Parental Abduction
10-82	Stalking
10-83	Infectious Disease
10-84	Demonstration

10-85	General Broadcast
10-87	Drugs
10-88	K-9 Unit Required
10-89	Tactical Unit Required
10-90	Bank Alarm
10-91	Armored Car
10-92	Person in Custody
10-93	Road Block
10-94	Operation Red Leaf
10-95	Traffic Complaint
10-96	Warrant Execution
10-97	DL Suspended
10-98	Marine Accident
10-99	Officer Update Status
10-100	Bomb Threat

Oregon State Police

12-1	In Service
12-2	Out of Service
12-3	Return to Office
12-4	Call Office by Phone
12-5	Repeat Message
12-6	Contact Complainant
12-7	Vehicle Registration
12-8	Vehicle Registration / Legal Owner
12-9	Check PUC Status
12-10	Check for License
12-10A	No Valid License
12-11	Description License
12-12	Unable to Copy
12-13	Stations and/or Cars Call
12-14	Relay to Station
12-15	Locate for Emergency
12-16	Motor Vehicle Accident
12-16A	Vehicle Accident, Fatal
12-16B	Vehicle Accident, Injuries, No Ambulance
12-17	Vehicle Accident, Ambulance Dispatched
12-18	Dispatch Ambulance
12-19	Dispatch Tow Vehicle
12-20	Check for Record /Wants
12-20A	Can Subject Hear?
12-21	No Record or Report
12-22	Prior MisdemeanorRecord - Not Wanted
12-23	Prior Felony Record - Not Wanted
12-24A	Subject Wanted - Felony
12-24B	Subject Wanted- Misdemeanor
12-25	Similar Subject Record, Added Info Required
12-26	Base Out of Service
12-27	Call by Radio
12-28	Suspicious Person
12-29	Disturbance
12-30	Reckless Driver
12-31	Intoxicated Driver
12-32	Intoxicated Person
12-33	Emergency
12-34	Resume Operations
12-35	Abandoned Vehicle
12-36	Illegal Hunting - Vicinity
12-37	Advise Road and WX
12-38	Switch Radio Frequency
12-39	Attention All Stations
12-40	Standby, Busy
12-41	Go Ahead
12-42	No Traffic
12-43	Disregard Previous
12-44	Accident or Spill - Hazardous Material
12-45	Burglar Alarm
12-46	What is the Telephone Number of Your Station?
12-47	Computer Files Unavailable
12-48	Computer Files Now Available
12-49	Death Investigation
12-49A	Possible Homicide
12-50	Message Not Radio Traffic, Handle by Phone

12-51	Radio Repairs at (Station No.) or Vehicle Number	19	Miscellaneous
		20O	DOT Advised
12-52	Radio Technicians En route to Your Station	2A	55 MPH Citation
		2B	65 MPH Citation
12-53	Regular Power Out, Using Emergency Power	2C	Seatbelt Citation
		2D	Childbelt Citation
12-54	Testing Station	3A	Seatbelt Warning
12-55	Transmit Equipment Test	4A	DUI Arrest - State Highway
12-56	No Help Available	4B	DUI Arrest
12-57	Disabled Motorist		
12-58	Narcotic Activity	**Orlando, Florida Sheriff's Dept.**	
12-59	Late Return	10-11	Dispatching too rapidly
12-65	Roll Call of All Units	10-12	Officials/Visitors
12-88	Off Duty	10-13	Weather/Road
12-94	All Clear - No Assistance	10-14	Convoy or escort
12-96	Vehicle Stop	10-15	Prisoner in custody
12-97	Radio Check Only	10-16	Pick up prisoners
12-98	Officer Needs Help - Non Emergency	10-17	Pick up papers
		10-18	Complete assignment quickly
12-99	Officer Needs Help - Emergency		
		10-18X	No lights/sirens
		10-19	Return to station
Disposition Codes:		10-20	Location
1	No Action Taken	10-21	Phone Call
2	Traffic Citation Issued	10-22	Disregard
3	Traffic Warning Issued	10-23	Stand by
4	Lodged In Jail	10-24	Trouble at Station
5	Report Taken	10-25	In contact with
6	No Report Taken	10-26	Did you receive message?
7	Log in Daily Report	10-27	Computer hit
8	Unable to Locate	10-28	Check vehicle registration
9	Field Identification	10-29	Stolen/Wanted
10	Unfounded	10-30	Against rules and regulations
11	Referred To Other		
12	Recontact	10-32	Is Breathalyzer operator available?
13	Civil Matter		
14	No Patrol Available	10-35	Confidential information
15	Follow Up	10-38	Block roads at_____
16	Information Obtained	10-39	Message delivered
17	Truck Inspection	10-42	Out of service at home
18	Cite And Release	10-43	Any traffic for_____

10-45	Call__by phone at__
10-50	Stopping vehicle
10-51	En route
10-52	ETA
10-54	Negative
10-56	Meet__at__
10-61	Open East/West Sallyport
10-62	Unable to copy
10-64	End of message/clear
10-66	Cancel
10-81	Eat
10-82	1-Person unit
10-88	Where can___be reached?
10-94	Give test count to 5
10-37	Operator on duty
10-96	Changing to Talkgroup___
10-97	Arrived
10-39	Message delivered

Dispatch Signals:

0	Armed \ use caution
1	Drunk driver
3	Hit and Run
4	Accident
5	Murder
6	Escaped prisoner
7	Deceased person
8	Missing person
9	Stolen tag
10	Stolen vehicle
11	Abandoned vehicle
12	Reckless driver
13	Suspicious vehicle
13P	Suspicious person
14	Information
15	Special detail
16	Obstruction on roadway
17	Contact
18	Felony
19	Misdemeanor
20	Mentally ill person
21	Breaking and Entering
22	Disturbance
23	Robbery
24	Investigation
25	Fire
26	Drowning
27	Prowler
28	Suicide
28A	Suicide attempt
29	Alarm
30	Rape
31	Approach on foot
33	Advise sergeant
37	Ambulance
38	Wrecker
43	Officer needs help
44	Officer needs assistance
51	Red light out
52	Caution light out
53	Green light out
54	Arrow light out
55	Bomb explosion
55A	Bomb threat
61	House check
66	Rape
66A	Rape attempt
88	Civil disturbance
95	Drug violation

Santa Barbara, CA Sheriff's Dept.

Code 1	Handle call at your convenience; acknowledge.
Code 2	Urgent, handle call immediately; no light or siren.
Code 3	Emergency,
Code 4	No further assistance
Code 4A	No further assistance needed; suspect not in custody.
Code 5	Stake out
Code 6	Out of car
Code 6C	Dangerous suspect. One-man unit stand by for assistance.

Code 7   Out of service to eat
Code 8   Fire call
Code 9   Vehicle stop at_____
Code 10   Out for warrant or subpoena
Code 12   Patrol your district and report extent of disaster damage
Code 13   Major disaster.
Code 14   Resume normal ops
Code 20   Notify press
Code 30   Burglar alarm ringing.
Code 33   Emergency traffic
Code 40   Traffic hazard exists
Code 50   Disturbance
Code 60   Traffic violation
Code 70   Service of vehicle
Code 80   Found property.
Code 91   Explosion.
Code 99   Emergency
Code 100 Bomb threat (or in position to intercept subject).
Code 1000 Begin operation

11 Codes:
11-4   Potential emergency.
11-5   Public relations
11-6   Discharging firearms.
11-7   Prowler
11-8   Person down.
11-10   Take a report.
11-12   Loose stock
11-13   Injured animal.
11-14   Animal bite.
11-15   Ball game in street.
11-17   Wires down.
11-24   Abandoned vehicle.
11-25   Traffic hazard.
11-25X Woman motorist.
11-26   Wants, warrants
11-27   Request driver's license
11-28   Vehicle registration
11-29   No record or current wants.
11-30   Incomplete call
11-31   Person calling for help.
11-40   Advise if ambulance is needed.
11-41   Ambulance is required.
11-42   Ambulance not required.
11-43   Ambulance followup
11-44   Coroner required.
11-45   Attempted suicide.
11-46   Death (non-traffic).
11-47   Injury (non-traffic).
11-48   Transportation request.
11-50   Field interview at_____
11-51   Security check
11-60   Investigate water leak.
11-65   Investigate signal light
11-66   Defective traffic device
11-70   Fire alarm
11-71   Fire report
11-79   Traffic accident
11-80   Traffic accident; major
11-81   Traffic accident; minor
11-82   Traffic accident; no injury.
11-83   Traffic accident; no details
11-84   Direct traffic.
11-85   Send tow truck.
11-86   Special detail.
11-98   Meet the officer
11-99   Officer needs help

900 Codes:
901   Ambulance call
901A   Ambulance call - attempted suicide
901B   Ambulance call - drowning
901C   Ambulance call - cutting
901D   Ambulance call - drunk

901G	Ambulance call - gas
901H	Ambulance call - dead body
901K	Ambulance has been dispatched
901L	Ambulance call - narcotic overdose
901N	Ambulance requested
901S	Ambulance call - shooting
901T	Ambulance call - traffic accident
901Y	Request ambulance if needed
902	Accident
902H	En route to hospital
902M	Medical aid needed
902T	Traffic accident
903	Aircraft crash
903T	Aircraft in trouble
904	Fire
904A	Fire alarm
904B	Brush fire
904C	Car fire
904I	Illegal burning
904S	Structure fire
905D	Dead animal
906K	Rescue Dispatched
907	Minor disturbance
907A	Loud radio or TV
910	Can you handle call?
911B	Contact informant
914	Request detectives
914A	Attempted suicide
914C	Request coroner

CYBER CRIME

How to Protect Yourself From Computer Criminals

Online harassment and stalking, security for your e-mail, data security, viruses, fraud, credit card access, children and pornography, threats to your privacy - there are dozens of evils out there, beyond your monitor screen. Here are simplified answers to protecting your computer, your family, yourself and your business from the cyber nasties cruising the Internet and the World Wide Web. *Cyber Crime* is a non-technical guide, written in plain English, which offers easy-to-implement answers and an indispensable appendix including a list of online resources and glossary. This Cook's tour of online crime is a quick 'n easy guide to staying clear of the dark side of cyberspace. You and your computer will rest a lot easier once you've read *Cyber Crime* and followed the suggestions it offers.

$16.95 plus $4.50 shipping/handling from LimeLight Books, P.O. Box 493, Lake Geneva WI 53147 Visa/Mastercard orders: 1-800-420-0579 (M-F, 9am-7pm EST). www.tiare.com/cyber.htm

INDUSTRIAL ESPIONAGE CAN KILL YOUR BUSINESS!

Corporate spying is much easier and far more widespread than it used to be. And it can kill your business. Kill it dead! 72% of businesses which have suffered a loss from corporate spying go out of business within two years! Even a whiff of a security breach can send a company's stock tumbling or kill a pending business deal. Protecting yourself from such a calamity can be an extremely costly process if you hire experts and/or go out and buy expensive professional equipment.

There's another way. *Improvised Technology in Counter-Intelligence Applications* by Dr. Robert Ing shows you how to use a common radio scanner and other inexpensive, easily obtained electronic gear to easily detect telephone taps, hidden radio transmitters, videocameras and tracking devices. Originally presented only in workshops the author gives to federal government personnel, this inside information is now available to businesses and corporate security people, private law enforcement, security and investigative personnel.

If you need to know whether your privacy is being compromised electronically and don't have the budget to buy highly specialized equipment or don't have the technical background to operate it - this manual will help you solve your security worries.

$29.95 plus $4.50 shipping/handling from: LimeLight Books, P.O. Box 493, Lake Geneva, WI 53147. Visa/MasterCard orders: 1-800-420-0579 (M-F, 9am-7pm EST). www.tiare.com/itech.htm

ON GUARD!

How You Can Win the War Against the Bad Guys - is a valuable, personal weapon in your war against the creeps and crazies out there. *ON GUARD's* "Crime Primers" give you vital information on everything from home security to carjacking and gangs, and will help you avoid becoming a victim. It's information that could even save your life or that of someone you love!

But *ON GUARD* is more! It's a complete guide to organizing your community to reduce - perhaps nearly eliminate crime in your neighborhood. *ON GUARD's* "Citizen Patrol" section shows you what to do and how to do it. Follow *ON GUARD's* recommendations and you and your neighbors will soon be serving as additional eyes and ears for the police. And that's the last thing a criminal wants — next to jail!

ON GUARD even includes a section of reproduceable "Crime Fighting Forms" you can use to record suspicious sightings and other information. These paper weapons can play a winning role in your personal war against the bad guys. (The nine forms are also available separately.)

If you're worried about crime you need *ON GUARD* and the practical guidelines and essential information it gives you on the best ways to stay as safe and free from crime as possible in the high-risk 90s!

$17.95 plus $4.50 shipping/handling from: LimeLight Books, P.O. Box 493, Lake Geneva, WI 53147. Visa/MasterCard orders: 1-800-420-0579 (M-F, 9am-7pm EST). www.tiare.com/onguard.htm

PAST DUE!

A Debt Collecting Manual

This extremely valuable manual shows you how to collect past due accounts using the same techniques as professional bill collectors. *Past Due!* tells you what you should say to debtors and what you should *never* say. Explains exactly how you should format your collection phone calls for maximum effect, what questions to ask the debtor how to handle their excuses and delays, legalities, you sample collection letters and much more!

Written for use in seminars the author gives for professional debt collectors, *Past Due!* is packed with powerful insider information explained by a master collections expert! Unleash the money you're owed by others. *Past Due!* is your ticket to a fatter bottom line!

$39.95 plus $4.50 shipping/handling from: LimeLight Books, P.O. Box 493, Lake Geneva, WI 53147. Visa/MasterCard orders: 1-800-420-0579 (M-F, 9am-7pm EST). www.tiare.com/pastdue.htm

**DON'T LET THE BAD GUYS WIN!
ORDER NOW!**

ORDER FORM

Name _____

Address _____

_____

CityState _____ Zip _____

Phone _____

Please send me:

_____ copies of **CYBER CRIME** at $16.95 each.

_____ copies of **IMPROVISED TECHNOLOGY** at $29.95 each.

_____ copies of **ON GUARD** at $17.95 each.

_____ copies of **PAST DUE!** at $39.95 each.

*Add $4.50 s/h first book, $1 for each additional book
or forms set ordered*

My ☐ check ☐ money order for _____ is enclosed.

(WI residents add 5% tax)

Credit Card ☐ Visa ☐ MasterCard

Card # _____ Expiration _____

Name as it appears on card _____

Signature _____

Detach this form and mail to:
TIARE PUBLICATIONS
PO Box 493
Lake Geneva, WI 53147

Telephone Orders

Call (800) 420-0579
Mon.-Fri. 8 a.m. - 6 p.m. CST